the Lady and the Sharks

Eugenie Clark

the Peppertree Press

Sarasota, Florida

ISBN: 978-1-936051-52-6
Library of Congress Number: 2010920926

Printed in the U.S.A.
Printed February 2010
Fourth Printing September 2012

Acknowledgements

MY GRATEFUL THANKS are due to the volunteers of Mote Marine Laboratory, who raised funds, arranged publishing, and spent much time editing and revising the manuscript for this updated 4th edition: Joan Dropkin initiated this new edition; Greg Hoffmann coordinated the teamwork; Joe Lappin and Barb Hulyk served as editors; Dave Harralson, Cathy Marine, Rachel Dreyer and Linda Pine provided additional editorial help; and Joan Dropkin, Domenic Ali, and Taylor Mayou who selected and prepared the new photographs. Additional assistance was provided by many other Mote staff, including Lawson Mitchell, Sue Stover, and Donna Basso.

Genie with Joan Dropkin, a volunteer at Mote and Genie's yoga teacher, who initiated this new edition.

Charles M. Breder, Jr. 1897-1983

Dedication

To Dr. Charles M. Breder, Jr. (1897-1983)

THROUGHOUT THIS BOOK, I refer to Dr. Breder, my mentor since college days, who advised me to start this Lab. I never took any major step in the administration of the Lab or my scientific career so closely bound with it, without seeking Dr. Breder's sound advice. And he always seemed to have time, right when I needed it, to discuss in his wise, stimulating and encouraging way, my problem – whether it be the mysterious gonad of a little grouper, the bell-ringing behavior of a shark, or moving the Lab from Placida to Sarasota. My talks with Dr. Breder are the most wonderful I have ever known. His widow, Pricilla, and I, like many of his other students, relish memories of conversations that influenced our lives and brought us so close to his brilliant mind.

Dr. Breder once approached a young visitor at our Lab looking at a fish in a jar. "What do you think that fish is, young feller?" Dr. Breder asked him. The boy told him authoritatively, "Well, according to Breder it's ---," and he took out his copy of the famous Field Book of Marine Fishes of the Atlantic Coast, a best seller that Breder whipped out for "boy scouts" during his long train commutes from his home in New Jersey to his office as curator of fishes in the American Museum of Natural History in New York where he conducted and published his scholarly ichthyological research papers and books. His field book is also on the shelf of every library and in the personal collection of every ichthyologist I know.

My children adored Dr. Breder, the twinkle in his eye and his unique sense of humor got to each of them. They took it for granted that he was a genius. Hera is still inspired by her early associations in the Lab and recently asked, "How many seven-year-olds got to work as an assistant to Dr. Breder?"

Dr. Breder was adored and greatly respected by so many students of all ages and by scientists, all over the world, concerned with fishes, their anatomy, design, and behaviors. When he chose to retire near our Laboratory, I was able to easily call on him for advice and discussions. What a privilege it has been to have his wise counsel and advice during the critical stages in the development of our Laboratory.

Foreword
By Sylvia A. Earle

"SHE LOOKS TO ME to be about fourteen years old, but she really seems to know what she's doing." That was my father's assessment after meeting Dr. Eugenie Clark at the newly-established Cape Haze Marine Laboratory in 1955. He had driven me from our home in Dunedin to Placida, Florida to meet the author of my treasured book, *The Lady With A Spear*, and both of us came away deeply impressed by the slim, dark-haired scientist whose second book, *The Lady and the Sharks*, is updated and reproduced here.

She asked that I call her "Genie," and a "genie" she proved to be, magically transforming the lives of those around her, including my own. I try to imagine how she was regarded as a student at Hunter College in New York, among her mostly male colleagues at Scripps Institution of Oceanography, by the Palauan spear-fishermen, chieftans, and the men and women of various Pacific islands who accepted her as a friend, and by the scientists at Egypt's Red Sea field station at a time when science was hardly considered an appropriate career choice for a woman, let alone science that involved diving into the ocean where few had ventured. Her deep love and respect for the ocean, for science, and for all living things, from people to fish to the smallest sea creature, won admiration and affection from colleagues around the world. To me, she was living proof that it was possible to craft a full life as a professional scientist, explorer, founder and laboratory director while successfully balancing other roles as wife, mother, and daughter. Genie's parents lived nearby and were a part of an extended household that sometimes included visiting biologists such as I.

The Gulf of Mexico became Genie's big blue laboratory, one that we shared as she began observing fish and other marine life and I began exploring and documenting the ecology of the vast meadows of marine algae and sea grasses along Florida's west coast. I basked in her presence during dives at Boca Grande, Sarasota's Point-of-Rocks, and the Dry Tortugas; learned while watching her mesmerize audiences with stories laced with new insights about the ways of fish and other marine life; and marveled at her intelligence, style and grace.

The story of how the Cape Haze Marine Laboratory morphed into the renowned Mote Marine Laboratory reads like a fictional fantasy adventure, but the sea stories in this updated volume are far more exciting, because they are, of course, true. They also provide a clear record of the way the ocean was—prior to the sharp decline of ocean wildlife brought about by destructive fishing and overconsumption of sharks, tunas, marlin, grouper, tarpon, and many other creatures; before coastal waters of Florida and elsewhere in the world were darkened by polluted "dead zones;" before anyone imagined that global warming, sea level rise, and ocean acidification could be matters of concern in their lifetime.

When I met Genie, sharks were abundant in the Gulf of Mexico and throughout most of the oceans of the world, and continued to be common when Perry Gilbert led the initial development of the Mote Marine Laboratory. Few imagined then that sharks could become scarce, that they and many other common ocean species could plummet by 90% in a few decades, that coral reefs everywhere would soon decline by half. The last of the once-common Caribbean monk seals was seen in the Gulf of Mexico in the 1950s, but at the time, it seemed that for most species, the ocean was infinite in its capacity to yield whatever people wanted to extract, from wild fish, shrimp and lobsters to oil, gas, sand and rocks, while accepting whatever was put there.

Among the great discoveries of the 20th century is the fundamental role of the ocean in providing life-support functions for the planet, driving climate and weather, regulating temperature, yielding most of the oxygen in the atmosphere, and absorbing much of the carbon dioxide.

The Shark Lady and Her Deepness

We also have discovered that humankind has the capacity to disrupt these basic processes. Now we know that there are limits to what can be taken out or put into the sea without causing trouble for the ocean, and therefore, trouble for us. It follows that taking care of the ocean means taking care of ourselves.

In the preface to a previous edition of *The Lady and the Sharks*, Genie wrote, "I have written this book for those who love the sea whether they dive in it for fun or exploration or simply are eager to know more about it." Knowing is the key. With knowing, comes caring, and with caring there is hope that the wild and wonderful ocean that Genie describes here will continue to have champions who will protect and restore health to the blue part of the planet. Thank you, Genie, for all that you have done—and are doing—to make it so.

Sylvia A Earle

Member of Mote Marine Laboratory Board of Trustees,
Explorer in Residence, National Geographic Society
January 2010

PROLOGUE:
THE BIRTH AND GROWTH OF MOTE MARINE LABORATORY

By Kumar Mahadevan

ONCE AGAIN, we undertake a new edition of Eugenie Clark's book with a great sense of pride and a warm feeling for the rich history of Mote Marine Laboratory. A special thanks and kudos to the team of Mote Volunteers for their generosity and untiring efforts towards this republication.

Fifty-five years ago when Genie became the first executive director of the Laboratory, the world of science was very different. The technological breakthroughs from cell phones to the internet that we have witnessed in the ensuing years are remarkable. In spite of the advancements, the contributions of Genie and other scientists who conducted their research at Cape Haze Marine Laboratory and Mote Marine Laboratory remain ageless. As her book so vividly describes, Genie began building an international recognition for the Laboratory with her research on sharks. The foundation that the Vanderbilts, Eugenie Clark, Charles Breder, Stewart Springer, Sylvia Earle, and many other scientists laid during the first twelve years (1955-67) of the Laboratory's existence was strong and helped build the reputation and the strength the Laboratory enjoys today.

Mr. and Mrs. William R. Mote, Mrs. Elizabeth Mote Rose, and Dr. Perry W. Gilbert built successfully on the strong scientific foundation that was laid in the early years. The continuing vitality that the resident scientists and visiting investigators provided for the next eleven years (1967-78) launched the Laboratory into a new era of scientific excellence. The excitement that Genie began never waned. From Placida to the new facilities on Siesta Key, not a beat was lost. With expanded facilities and a greater recognition, Perry and Mr. Mote diversified into many research areas including red tide, biomedicine and environmental health, which had become critical concerns for Floridians then, and still are today.

Thanks to the tireless efforts of Bob Johnson, Dr. Perry Gilbert and Mr. Mote, the Laboratory moved to City Island, Sarasota in 1978, and a new era began. With an influx of new resident scientists and a combination of distinguished scientists such as Perry Gilbert and Bill Tavolga, and with a new director, Dr. William H. Taft (President, 1978-83), the Laboratory blossomed into a full fledged marine laboratory conducting research in the areas of shark biology, biomedical research, marine mammals and sea turtles, aquaculture, coastal resources, chemical fate and effects, and environmental assessment and enhancement. During those five years, the annual budget grew from $300,000 to $2.2 million (today, our annual budget is about $20 million). While maintaining its leadership in marine science research throughout the previous three decades, the Laboratory became involved in educating youth (Jim Wharton currently leads this effort) and supporting the community in various environmental protection and enhancement activities as well. Thanks to the leadership of Dan Bebak, Mote Aquarium has grown to become one of the most popular attractions in southwest Florida while proudly showcasing the research currently underway at the Laboratory. And in order to provide a greater societal impact, the Marine Policy Institute was created in 2006 and, with the leadership of Frank Alcock and Barbara Lausche, is bridging science with policy on important environmental issues such as red tide and sea level rise.

I am very proud of my association with the Laboratory as a senior scientist for the past thirty-one years and as its CEO for twenty-three years. It is particularly gratifying to enjoy the strong partnership of the past with the present. Eugenie Clark, Bill Tavolga, and many other scientists who have built the excellent scientific reputation of the Laboratory continue to stay involved as supporters and active researchers, while today's scientific leaders, Ernie Estevez (coastal ecology), Bob Hueter (sharks), Ken Leber (fisheries), Kevan Main (aquaculture), John Reynolds (marine mammals), Rich Pierce (ecotoxicology) and David Vaughan (coral reef), working with a great team of researchers (Aaron Adams, Erich Bartels, Jim Culter, Kellie Dixon, Mike Henry, Barbara Kirkpatrick, Gary Kirkpatrick, Carl Luer, Jim Michaels, Kim Ritchie, Tony Tucker, Cathy Walsh, Carl Walters, Randy Wells, Dana Wetzel and others) are continuously enhancing the scientific stature of the Laboratory. Our recent affiliation with the University of South Florida has strengthened the research partnerships between the two organizations and further enhances the scientific reputation of the Laboratory. Since our move to City Island in 1978, several wonderful chairmen of the Board of Trustees have helped the Laboratory to prosper: through 1986 William R. Mote; 1986-90 Bob Johnson; 1990-93 Richard Angelotti; 1993-96 Michael Martin; 1996-99 Alfred Goldstein; 1999-2002 Frederick M. Derr; 2002-04 Myra Monfort Runyan; 2004-07 Mike B. McKee; 2007-09 Judy Graham; 2009-present Arthur L. Armitage.

Many volunteers and benefactors have continued their enthusiasm and support for the Laboratory. All of these make Mote Marine Laboratory a great organization. As it was 55 years ago, research freedom for our scientists continues to thrive. What Genie gave birth to has grown up to be an outstanding citizen. We are proud of her and are especially grateful for her allowing us to reprint this wonderful book of history, tradition and excitement. And, it is all the more special since the book is rededicated to one of the greatest ichthyologists of our time, Dr. Charles M. Breder, Jr.

Personally, Genie has been a mentor, friend and a wonderful cheerleader. I have benefited tremendously over the years from her

Directors of the Lab Front Row: Dr. Eugenie Clark, 1955-1965, Dr. Perry W. Gilbert (1912-2000) 1967-1978; William R. Mote (1906-2000). Back Row: Dr. Kumar Mahadevan, 1986-present, Dr. Sylvia Earle, Interim Director, 1966

wise and always optimistic counsel. As my friend "Her Deepness" Sylvia Earle so eloquently described in her foreword, Genie's enthusiasm, energy, radiant happiness, and unending curiosity are contagious and have spread throughout Mote Marine Laboratory. Genie is "Rachel Carson" for many of us and has meant much to our careers.

Kumar Mahadevan, Ph.D.

President/CEO
Mote Marine Laboratory
January 1, 2010

Contents

1

Discovering
a Watery Eden

THE WEST COAST OF FLORIDA is edged by a series of sandy keys, with bays between them and the mainland, and the Gulf of Mexico on their outer sides. Where man has not cleared the bay shoreline, it is dense with mangrove trees whose overhanging branches and roots, reaching into the water, provide protective cover and substrate for the rich marine and estuarine life. If you look at a map of Florida, two huge drainages stand out on the west coast, Tampa Bay and Charlotte Harbor. From these the rains and tides wash abundant nutrients into the Gulf. Only Charlotte Harbor, of the large estuaries in this country on the Gulf of Mexico and Atlantic Coast south of

Cape Cod, is still virtually unpolluted. Deep Boca Grande Pass, the entrance to Charlotte Harbor, is the most famous place in the world for tarpon fishing and large hammerhead sharks—a real challenge in bringing your tarpon boatside in one piece. Northwest of Boca Grande, Charlotte Harbor leads into Gasparilla Sound, then narrows to a slit which opens into Lemon Bay. Big and Little Gasparilla islands are said to be the place in which the pirate Gasparilla buried his treasure.

My introduction to the west coast of Florida was in 1954 after I accepted an invitation from Anne and William H. Vanderbilt to give a lecture in Englewood, Florida. Mrs. Vanderbilt had read my book *Lady with a Spear* and talked her husband into reading it. Their ten-year-old son, Bill Jr., had a bedroom full of aquariums, as I did at his age, and his parents had become fascinated with their son's hobby. Their estate stretched across Manasota Key, from Lemon Bay to the Gulf of Mexico. Bill Jr. and his school chums, like all children living near the water, explored the shore and brought home all kinds of strange sea life they found in shallow water or washed up on the beach. But many of these items they couldn't identify; a few mystified them even after they consulted books, teachers, and the local fishermen. There was no marine biologist in the area.

William and his brother, Alfred Gwynne Vanderbilt, had bought a 36,000-acre tract of land near the fishing village of Placida, southeast of Englewood, extending onto the Cape Haze peninsula. They had started raising Santa Gertrudis cattle at their newly formed 2V Ranch and were developing land along the miles of waterfront property in this beautiful, isolated, semitropical country. Alfred Vanderbilt had also decided to build a house in their Cape Haze development, where his family could enjoy the beaches, boating,

2

and fishing and where, in his private plane, he could easily hop over to see his horses when they ran at Hialeah.

My evening talk was given in the Englewood public school and was open to the public. There were many children and commercial fishermen with their families. Some stood in the doorways to hear what I thought would have limited appeal— a lecture on Red Sea fishes. I learned from the comments as I lectured that for almost every unusual fish I mentioned or showed on colored slides, a similar fish had been seen at Englewood; giant sting rays, mantas, guitarfish, scorpion fish, electric rays, nurse, hammerhead, tiger sharks, and many other spectacular fish were known also to the local fishermen, who wanted to find out more about them in the long question/answer session after the lecture.

My own interest in sea life had begun when I was in elementary school in New York. My American father died when I was a baby, and my Japanese-born mother was working at the cigar and newspaper stand in the lobby of the Downtown Athletic Club. On Saturdays, while she worked, she left me nearby in the old New York Aquarium at Battery Park, where I spent many hours watching the fishes. Afterward we usually went to eat at a charming little Japanese restaurant, Fuji, and gradually became good friends with the owner and cook, Masatomo Nobu, who later became my stepfather. I was brought up on the Japanese side of my family, but no one, except anthropologists who are quick to spot my Mongolian eyefold, ever thinks I'm part Japanese.

I knew more about produce from the sea than any of my schoolmates, and my reports in school, from kindergarten on, amused and shocked my classmates and teachers. I told them how we ate with chopsticks, had rice and seaweed for breakfast, raw

3

fish, octopus, and sea urchin eggs for supper, and cakes made from sharks. I was the only student of Japanese ancestry in the schools where I grew up, in Woodside, Long Island.

Nobusan often visited my family (Grandma Yuriko, Uncle Boya, and Mama) and always brought us some special Japanese delicacies from his restaurant. He already seemed part of the family when he became my stepfather, at the time I graduated from Hunter College. I had majored in zoology. Since my first visits to the Aquarium at Battery Park, I had wanted to be an ichthyologist. Graduate courses at the University of Michigan, Columbia University, Woods Hole, Scripps Institute of Oceanography, and at New York University, where I got my master's and doctorate degrees, prepared me for scholarships that took me to various parts of the world to study fishes. I had visited marine laboratories in many countries, and in my book I described a small, isolated marine biological laboratory in the little fishing town of Ghardaqa, on the Egyptian coast of the Red Sea, where I had studied fishes for a year as a Fulbright Research Scholar in 1951.

I never dreamed I would have the opportunity to start a laboratory from scratch myself, but later on the evening of my lecture in Englewood I learned that the Vanderbilts had invited me to Florida for just this reason. They felt the stimulating, strong interest and importance of marine life in their part of Florida, and wanted to help develop this interest in some worthwhile way. Bill Vanderbilt told me, "It's Anne's idea. She thinks it would be just great if we had a marine laboratory here, something like the one you described in your book. The rest of us are all for it, too. What do you think? Is this a suitable place for that kind of research? Would you consider starting a marine laboratory and being the director?"

It was an exciting offer—an opportunity to study fishes and other marine life and to create a laboratory. The west coast of Florida was quite different from the tropical coral reef areas where I had done most of my research. There were many basic things I would have to learn about the less clear waters around a major estuary. And there were no strings attached, no contract; I didn't have to make any promises, just "Start a place here where people can learn more about the sea" was all the Vanderbilts asked of me.

I knew there were no marine biological laboratories in this part of Florida. In the 1930s, the New York Aquarium had had a small research station some miles south on Palmetto Key. In my studies I had often referred to the valuable scientific publications of Director Charles M. Breder Jr. and other scientists who had studied how tarpon got into fresh water, the mouth-breeding habits of poisonous catfish, the floating leaf mimicry of baby spadefish, and other fascinating phenomena of sea creatures. This series of scientific studies seemed to be just the start of an endless amount of research that could be done on the fertile, complex part of the sea in the estuaries and on the wide, rich continental shelf in this part of the west coast of Florida.

Unfortunately, this kind of basic research does not often get the support that applied research can get. The Palmetto Key Laboratory closed after a few years for lack of funds. Many a scientist has had to leave the freedom, creativity, and exciting experimental studies in basic research for restricted research in an industrial or other "practical" laboratory in order to support a growing family. It is hard for many people to understand the basic scientist's desire to learn the answer to a question just for the sake of new knowledge. Not many appreciate the ultimate power and potential usefulness of basic knowledge accumulated by obscure investigators who, in a lifetime of intensive study, may never see any practical use for their findings but who go on seeking answers to the unknown without thought of financial or practical gain.

Some scientists simply prefer to work in a field of abstract mathematics with as yet no practical application, or to look for the reason a pea plant is tall or short, or the color of a fish can turn from yellow to blue, or the fruit fly's eye is red or white. They are simply interested in learning why and how things work for sheer curiosity's sake. Such scientists often become characters burlesqued as "absent-minded professors" with little money, who don't care how they dress, and who forget where they leave their glasses. Yet this is the type of man who, because of his years of study and understanding of the implications of a problem, returns to the peace of his laboratory to examine his ten thousand and first platyfish and discovers the mechanism for the inheritance of a black tail pattern.

Sometimes a basic research scientist does not finish his studies in his lifetime. Sometimes many lifetimes are put in by many investigators before one person works out the last part of a problem. Years later, it may prove to have an unforeseen and wonderful practical application; or it may just rest—an exquisite piece of knowledge appreciated by only a few specialists or people with vision.

It is not usual to meet people with the Vanderbilts' background who have a natural understanding of why marine biologists in basic research seek answers to what many consider trivial questions. Anne, Bill, and Alfred were as interested in learning how a male pipe fish gives birth to babies as in any new trend in the stock market— perhaps more so. It's always a pleasure to talk about my work with people who share my curiosity about the sea without asking such questions as, "All that's very interesting, but what good is studying the difference between plectognath fish in the Red Sea and the Indian Ocean?" If I point out that some of these fish are poisonous and people die every year from eating the wrong kind, then my studies make some sense. Yet the poisonous quality of plectognath fish is not why I started studying them. I was intrigued with the puffing mechanism of blowfish, the box around the coffin fish, the still-unanswered question of why the three-toothed *Triodon* has a gigantic folding skirt on its belly. If, from my research, mankind gains some practical application or benefit, this is added delight and satisfaction to my work. But this is not what drives me to study late into the night or to watch a fish on the sea bottom making some strange maneuver until all the air in my scuba tank is gone and I hold my breath for those last few seconds of observation.

Anne and Bill Vanderbilt are a handsome couple, tall, dignified, yet with an easy manner, a sense of humor, and an enthusiasm for life and people. I could imagine William Vanderbilt looking as natural and at ease being inaugurated Governor of Rhode Island as he did in a cowboy hat and ranch clothes when he and Bill Jr. drove me around Placida. They showed me the beaches, bays, their ranch, Cape Haze, and finally a beautiful spot on Gasparilla Sound where they hoped a marine laboratory could be built.

Across the wide expanse of calm water Bill pointed out Big and Little Gasparilla islands, smaller mangrove islands, and Gasparilla Pass—the nearby exit for a boat from the Sound out into the Gulf of Mexico. Thousands of orange-and-purple fiddler crabs swarmed away from our feet and over mangrove roots as we walked along the edge of the water where Bill suggested we build the dock. A mullet leaped over a foot out of the murky water. Closer to shore, the smooth surface rippled here and there, stirred by schools of small fish. Here the water was shallow and clear enough to see to the bottom, where blue crabs stretched out their claws as our shadows reached them. I was going to take off my shoes and wade into the water, but changed my mind when I saw beds of oysters with sharp shells and spiny sea urchins, trying to hide themselves by carrying bits of shells on their backs as they crawled slowly over a bottom riddled with holes. Other parts of the shallow water had masses of tangled seaweed—green, brown, red algae with decaying blades of *Thalassia* indicating nearby grass beds.

Bill and Anne Vanderbilt had arranged to have a local fisherman take me seining. "You'll like Beryl," they told me. "He knows so much about the fish around here and how to catch them. And he's such a wonderful person."

Beryl Chadwick smiled easily. He was a thin, taut man, casual, modest, and a bit shy. His skin, especially on the back of his neck and his arms, was ruddy and a little leathery. His hands were calloused and strong. If you didn't ask him questions, you couldn't guess the wealth of knowledge and experience in this quiet man. When I first shook his hand, I had no idea how much I was to learn from him.

I changed into a bathing suit and wore sneakers to help Beryl pull his long seine. Anne came along "just to watch," but as Beryl and I pulled up sea horses, pipefish, young pompano, and the black leaflike babies of the gold-and-blue "foolfish,"* Anne couldn't see enough from where she stood on dry shore and she came to the

---

* The local fisherman's name for the scrawny filefish in the genus *Alutera*—which grow to about two feet long and can be caught easily with a dip net or sometimes even by hand.

edge of the water, her high-heeled pumps sinking into mud as she and Bill Jr. excitedly helped sort our catch. We got some formalin from the drugstore and preserved our fish collection. I warned Bill Jr. and his friends that concentrated formalin (embalming fluid) burns the skin.†

Beryl, besides being a commercial fisherman for many years, had worked at the Bass Biological Station in Englewood. I met young Johnny Bass and his wife Barbara later that day. They volunteered to help and guide me if I started a marine laboratory. One could easily imagine Johnny, who was strong and six feet tall, and his petite jeep-driving wife bringing back an alligator if we needed one. Like the Vanderbilts, the Basses also recommended Beryl to assist with a new laboratory.

Johnny's father had operated one of the largest biological supply houses in the country right there in Englewood. Hundreds of thousands of biology students must have looked under a microscope at preserved *amphioxus* (the classical example of a fishlike ancestor of the vertebrates, studied in every college) that started life in Englewood's Lemon Bay. The Bass Station had flourished because of the rich animal life in the area. The catalogue and price list of animals sold by the Station gave me an idea of the great variety and abundance of animals there.

Unfortunately, the Station closed after Johnny's father died, in 1938, when Johnny was still a child. But it left an impression

---

† It should be diluted ten times (1 part formalin to 9 parts water) as a preservative for fishes. Fish over two inches long should be injected or a knife slit made into the abdominal cavity so the formalin can quickly reach and preserve the viscera (stomach, liver, intestines, etc.). For tiny fish, worms, and jellyfish, a much weaker solution can be used. After specimens are preserved for several days, they can be washed and placed in a more pleasant-smelling solution such as 70% ethyl or 40% isopropyl alcohol.

on Johnny that would stay with him for the rest of his life and be transmitted to his children. Luckily, Johnny found a wife who also liked to go out to sea or into the mangroves or the Everglades jungle to catch and bring back animals. They were eager to see a biological laboratory set up again and to help us collect specimens.

The timing of the Vanderbilts' offer corresponded exactly with the completion by my husband, Ilias, of his medical training in New York City. He liked the idea of starting his practice of medicine in Florida and of bringing up our young daughter Hera and an expected second child in a beach house in Florida rather than continuing to live in our twelfth-floor apartment on West End Avenue. Even my mother and stepfather (who now ran a larger restaurant at Broadway and 124th Street) liked the idea. "Maybe we'll move Chidori to Englewood. Florida doesn't have one Japanese restaurant!"

In early January, 1955, six months after my first visit to Englewood, our family arrived in Florida. I opened the Cape Haze Marine Laboratory as soon as I arranged for a baby-sitter for two-year-old Hera and her month-old sister, Aya.

The Vanderbilts had prepared for me a small wooden building, 12 by 20 feet, with a sink and shelves for specimens. The building was constructed on skids, so that it could be moved if its location on the beautiful and undeveloped shore of Gasparilla Sound might not prove suitable. They had also built a dock, and Alfred Vanderbilt, who had a house nearby, gave the Laboratory the use of his 21-foot Chris-Craft, *Dancer*, named after his famous horse, Native Dancer. Beryl had collected more fish from Lemon Bay and put them in formalin in glass jars on a shelf. I knew my first basic job would be to begin identifying the local fishes and ultimately to publish my findings about them.

Then, the day after our arrival, I received a phone call from Dr. John H. Heller, Director of the New England Institute for Medical Research. He was in the Caribbean, in "shark-infested waters," looking for shark livers for his research, but he couldn't find a shark. He had heard that we were starting a marine laboratory. Could we get him a shark? I was interested in sharks, too, because of long-standing curiosity about the function of abdominal pores, which are common only to sharks and other primitive fishes. I put my hand over the phone. "Beryl, do you know how to catch a shark?" He straightened up, almost indignantly, "Stew Springer and I supplied the old Bass Station with all the sharks they needed," he said. Beryl indeed knew the ways of shark fishermen and set to work making a shark line.

Before Beryl and I had unpacked the supplies I'd ordered for our little laboratory, we were in the shark-hunting business.

Dr. Heller and his wife, Terry, came to Placida on January 24 to help catch their shark. Beryl's well-calloused hands showed the effects of hours of splicing rope, but the shark line was ready in time. Along three hundred feet of 3/4-inch manila rope he had spliced in 16 loops. Into each loop he attached, with a modified, quick-release becket knot, a side line and hook. Each side line had 6 feet of rope, then 3 feet of steel chain, and then a large 2 1/4-inch steel hook. On each end of the main line, Beryl attached an anchor and a 40-foot line with a float. At one end he rigged the float with a high flag marked "CHML." Beryl loaded the whole shark line in the *Dancer* and set the line two miles offshore in the Gulf. He baited each hook with a mullet.

Dr. and Mrs. Heller, Beryl, and I went out the next morning to check the line. Beryl brought along a bucket full of a fresh replacement of mullet. As the *Dancer* sped out toward the white flag, we could see the second float coming from the other end of the line about three hundred feet away, and knew the anchors had held the line properly taut. We brought the boat alongside the flag. Beryl pulled up the float and started working down the long line.

The first hook was empty.

"Did a shark take the mullet?" I asked him anxiously.

"I doubt it—maybe the crabs nibbled it off."

The Hellers looked at me and I guess they, too, were wondering how Beryl could tell if a shark had taken the bait or not. Beryl rebaited the hook and dropped it back in the water and continued along the main line. The second hook was empty. The third hook was empty also, but Beryl paused and made us look more closely. The side line and chain were twisted around and around. The metal hook, the thickness of a pencil, was partly straightened. "Wow!" was John Heller's only comment.

We leaned over the side of the boat, more anxious than ever, and watched the line as Beryl continued to pull it up and rebait the hooks. After a few more hooks, we saw Beryl straining to pull up the next

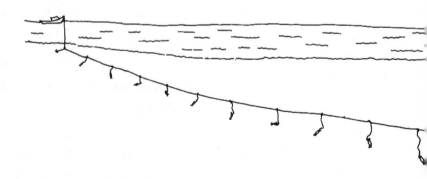

side line. John grabbed the line to help, and as the two men pulled, Terry Heller and I saw what the men felt—the line jumped. Several feet below, a large white object was coming up through the water. The chain led into the mouth on the underside of a shark. It was still alive and as the jaws moved we could plainly see the triangular serrated teeth. The eyes stared, but as the rope brushed against one side of the head the nictitating membrane blinked shut over one eye. The shark began to thrash about, showing its brownish-gray upper parts, and we saw that it was about eleven feet long.

The Hellers and I were thrilled and amazed, and could hardly believe our catch. Beryl beamed happily. "Well, what did you expect on a shark line!" He pulled the modified becket knot and released the side line with the shark from the main line, He tied the shark to a cleat on the side of the boat. Several hooks farther on, a smaller shark, about seven feet long, came up on the line. We towed back two fish totaling over 700 pounds. Terry Heller took movies of the sharks riding the stern waves in the sunlit foam.

We had no trouble rounding up some men—including Bill Vanderbilt, whose office was nearby—to help pull the sharks ashore

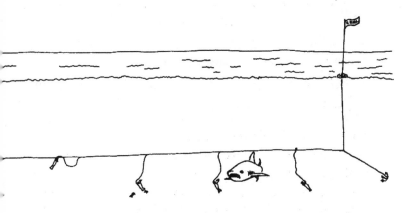

next to the Laboratory where we could dissect and study our specimens.

The dissection of a 500-pound shark, especially for the first time, is an exciting experience even for an ichthyologist who has studied the innards of many fish. Dr. Heller had never dissected a large shark before, and he couldn't wait to cut it open. Beryl handed him a big sharp knife.

John Heller, a medical surgeon of no small talent and experience, didn't know how tough the dermal denticles (modified teethlike scales covering the shark's body) were on a shark of this size. He had a difficult time getting the knife through. (Don't be deceived by the swimmer in the movies who grabs a large shark and easily slits open its belly with one swoop of the knife.) Beryl showed John how to cut from under the skin outward once the initial stab was made. Then John slit open the entire length of the belly and a huge gray fatty liver, about a fifth the weight of the whole shark and the full length of the belly, came sliding out. The stomach, full of water and partly digested chunks of fish, followed.

The stomach of a shark has two parts. Just after the esophagus (the pipe connecting the mouth with the stomach) is a huge, easily stretched "cardiac stomach," which, in the shark we opened, could hold half a human with ease. This narrows down abruptly to an equally long but very tight "pyloric stomach," which is more selective about what material passes through it before entering the intestine. It would barely admit one finger, and in it was only the well-digested goo biologists call "chyme." Sharks have a peculiar way of getting rid of large pieces of indigestible material from the cardiac stomach. They vomit (less explosively than we do)—an amazing phenomenon we were to witness later: The teeth are retracted in the soft flesh of the gums, the entire cardiac stomach merely everts out of the mouth, and the stomach lining is rinsed in sea water and gently slides back into place in much the same way one might turn inside out and rinse the pocket of a young boy's blue jeans in which a fleshy treasure had been forgotten for a few days.

The big dusky shark we were examining appeared well dead by now. Even the tail had stopped slapping the shallow water next to the beach where we had dragged out most of the body with the help of six men pulling the hook line. John cut open the pericardial (heart) chamber just anterior to the big belly, where we were examining all kinds of interesting organs in the abdominal or perivisceral cavity. The heart was still beating weakly. John massaged it and it pulsated faster. He started cutting into the floor of the mouth, touched the rows of sharp teeth, then discovered some of the blood oozing about was coming from one of his fingers. The big husky surgeon's voice became like a little boy's as he turned to his wife, the mother of his three children, and said, "Hey, Ma, look at my fingers."

Terry Heller, small and blond, was by now probing around in the viscera of the shark's abdomen. "Oh, go bleed someplace else, John!" she said. "Genie, come here. Are all these green blobs in the liver gall bladders?" Then Terry looked up and struggled to disenchant herself from the shark's viscera long enough to put a Band-Aid on John's finger.

The big purple spleen and pale pinkish-white pancreas, in several sections, looked similar to those in any other vertebrate. This dusky shark didn't have the classic "spiral valve" intestine that anatomy students study in the dogfish. The intestine was the "scroll" type, with a roll of membrane inside like yard goods, which served the same purpose, of giving a large surface for absorption in the

15

short intestine.

We were pleased to find the long rectal gland at the posterior end of the intestine. This interesting gland, whose function was unknown until a few years ago, is found only in elasmobranch fishes: sharks, rays, skates, sawfishes, and chimaeras. Elasmobranchs store dissolved urea salts in their bodies—an aid in keeping the body fluids close to the same osmotic pressure as that of sea water. True bony fishes living in the sea struggle with an osmotic imbalance, as they have "fresh water" in their body fluids. One Navy survival manual informs us that if you are adrift at sea you may save your life with drinking water by catching fish and sucking the fluid from the meat. But this won't work with sharks and other elasmobranchs; drinking their body fluids is not much better than drinking sea water. The rectal gland aids in keeping a better osmotic balance by retaining salt.

We continued to probe deeper into the viscera of the dusky shark lying on its back. The two big lobes of the uterus looked well stretched, and "placental" scars on the lining looked as if the shark had recently given birth to about a dozen young in the two compartments.

Lying close to the back "bone," or vertebral column, which in a shark of this size is composed of calcified units of cartilage but not true bone, we found the long paired kidneys. The ducts from the kidneys expanded near their ends to form a functional urinary bladder (which anatomy books tell you sharks don't have), and urine is excreted from a small penis-like organ, the urinary papilla, located between a pair of abdominal pores—the mysterious openings I had wanted to study since I first examined a small preserved shark at college.

By the end of the week, we had caught a total of twelve adult sharks, mostly dusky sharks and sandbar sharks. We had performed "Caesarean" operations and delivered live young from the uteri of both dusky and sandbar sharks, with Bill Vanderbilt and his Cape Haze Corporation office crew acting as midwives on one afternoon. Unfortunately, the embryos were premature. I tried to keep some in improvised plastic-bag incubators with aerated running water, but none lived for more than a week. Elmer Peckham, the caretaker of the grounds of the Cape Haze Corporation, thought that the tons of shark meat would make fine fertilizer for all the trees and bushes he was planting, but he gave up and let Beryl tow the carcasses out to sea when he realized how many man-hours it took to chop up a shark and bury the pieces around his plants.

I began keeping records on the sharks, adults and embryos. I was using a large handy reference book by Drs. Bigelow and Schroeder on sharks of the western North Atlantic. I was surprised at how few accurate and detailed measurements and dissection data were available on large sharks. The Lab lacked a means to measure accurately the weight of a large shark, so we decided to invest in a big meat scale. Beryl built a stand for it and a strong hoist at the end of the dock.

One day, the floats on the opposite ends of the shark line were floating close together when our boat got there. We had a mess of tangled lines, chains, hooks, and sharks. A school of dusky sharks must have passed over our lines, and six monsters, each between 10 and 12 feet long, each weighing over 400 pounds, had pulled the anchors loose.

All six were females and two were pregnant. Hera was at the Lab that day. As we removed the dying, soft-skinned, streamlined premature embryos, she cradled one that was still alive, holding the "placental" cord closed so it wouldn't lose any more blood, rocking it gently and humming to it as if she had a doll. Some of the embryos were dead but very fresh and clean-looking. That night I didn't let the babysitter or the children know that the tasty, boneless fish I fried for supper was filet of unborn dusky shark.

The first sharks Beryl caught were dying by the time we got them back to the Lab. They were weakened by their struggles on the line, and towing them behind the boat got water into their stomachs. Gallons of sea water would pour out when we cut open a stomach to find out what else the shark had been eating besides our mullet. We tried bringing one large tiger shark back by running the boat as slowly as possible and swimming the shark. Sometimes the shark got ahead of the boat and pulled us in another direction; an observer from shore might wonder who was leading whom. When we finally got a shark back to the Lab alive and in good condition, there was no place to keep it.

Dr. Heller explained that live sharks would be more useful for his cancer research, and I realized the same would be so for my investigation of the function of abdominal pores. He suggested we build a place where we could keep sharks alive. "You could study their behavior and the function of the abdominal pores, and I could try some experiments to see if a shark can synthesize radioactive squalene for our cholesterol research. A shark should be able to make the stuff in a few days after we inject it with C-14. Then we could sacrifice the shark and recover the tagged squalene from the liver." It sounded pretty simple. I could tie the pores closed while the shark was anesthetized and Dr. Heller was making his injection. In a few days, I hoped, I'd have the answer to an old question in my mind.

The Hellers' first visit was for only a week. But before they left we talked over with Beryl, and finally with Bill Vanderbilt, the construction of a 40- by 70-foot stockaded pen, adjacent to the Lab's dock, for holding sharks and other large fish. The Vanderbilts were pleased that the work during the first month of the Lab's existence was becoming part of Dr. Heller's medical research into cancer and cholesterol formation and that I was going to start experiments with sharks. The announcement of the opening of the Lab in several scientific journals also brought letters from scientists around the country asking to visit us and to use the Lab facilities as headquarters for collecting and making field studies. In the next few months, while the pen was being built and an additional room was added to the Lab, we received requests from two scientists who wanted to collect algae, two who wanted to hunt snakes on mangrove islands, and a zoologist-photographer who wanted to take photographs of the eyes of fish. Professor Bradley from Cornell came to collect mollusks and left his collecting gear, dredge, nets, and books as a contribution toward all the equipment the new Lab needed. Teachers and professors from nearby schools and the University of Florida asked if they could bring students to visit us.

We were building a sizable reference collection of preserved fish, crabs, shrimps, worms, starfish, and all kinds of other marine invertebrates. Barbara and Johnny Bass came over several times a week with at least one bucket of sea water in which some odd creature from under their dock was creeping, slithering, or swimming.

Johnny and his friend Clint Hancock from the Everglades caught a five-foot alligator in a mudhole across the road from the Lab. The Basses had a son, John, who was Hera's age, and young John and Hera were with us when Clint croaked down the mudhole, luring out the alligator which he and Johnny then caught with their hands. Beryl fenced in a section of the Lab's ground and made a cement pond, under the shade of a palmetto tree, for the alligator.

The local people brought us more alligators, snakes, turtles, a pelican with a broken wing. It was hard to keep the Lab from turning into a general zoo. I occasionally brought Aya to spend the day at the Lab, and she was perfectly happy if I set up her playpen next to an aquarium, the snake cages, or the alligator pen. Hera loved to play with the beautiful six-foot indigo snake Beryl found sunning on a bridge, and she would climb into the playpen with it and join Aya, who also loved this gentle Florida snake.

We had to obtain a permit from the State Board of Conservation to collect and keep some of the live animals. When the pen was ready for its first occupant, the Basses arrived with a large loggerhead turtle on the back of their jeep and Barbara had its eggs in a box on her lap. They drove the jeep onto the wide end of the dock where Beryl and Johnny unloaded the turtle, Miss Cape Haze, into the pen. We reburied her eggs in the sand near the pen, and Beryl built a wire cage over the area so the raccoons wouldn't dig them up. Barbara and Johnny had found the turtle the night before, crawling out of the water and up onto the beach, and had waited until she laid her eggs in the sand before they captured her for the Lab.

Beryl caught a tiger shark and a nurse shark of a reddish color, and these he named Hazel and Rosy. To this trio in our pen, we soon added two spotted eagle rays and a tarpon that Beryl got from the stop-net fishermen who set their nets in Gasparilla Sound near the Lab. Beryl had ordered a 16- foot flat-bottom skiff built with a large live well in the center. When he saw the fishermen closing their long net, he would run the skiff, with an outboard motor, out to join them and bring back any "trash" fish they didn't want. Some we kept alive, some we preserved. As I put labels on each fish in our reference collection, I had to refer to lots of books and scientific articles to identify each down to species. Beryl knew them all, of course, and in the records we kept he told me the local fishermen's name for each fish whose scientific name I recorded.

One day, the Basses came over and found me looking over my books trying to identify an odd jellyfish that Beryl had found floating in the bay.

"I think we've still got some books in my father's lab, Do you want to look them over?" Johnny asked.

We drove over to the old buildings of the Bass Biological Station and opened doors and drawers that hadn't been touched in years. Under cobwebs and dust we found dozens of valuable books and series of scientific journals, and publications on fishes, jellyfishes, worms, etc. sent to Johnny's father by scientist friends from all over the world. Beryl and I had to come back another day to haul away all the items we could use at the Lab that the Basses offered us. There was laboratory glassware, many unopened bottles of chemicals, and various instruments of the kind useful at a marine laboratory. Johnny's mother also was pleased that the Lab could use this material, and she asked if I wanted to glance through the records her husband had kept. I began to get a good picture of how the Bass Biological Station had operated and how useful it had been to scientists, teachers, and students. There were times, in those first months after we opened the Lab, when I wondered whether starting a marine laboratory here was really justified. Now I was sure it was, but I wasn't sure if, with my laissez-faire attitude toward the way I ran much of my life, I could give the Cape Haze Marine Laboratory the direction it needed to become a useful and respected organization, the way the Bass Station was, and not just a place where I satisfied my curiosity and need to study fishes. But there wasn't time to worry too much about this, and I rationalized that it would take me nearly a year to get the feel of things before I set down any definite policies or programs for the Lab other than our general commitment to study the local marine life.

It did not take Beryl long to find out I loved to dive, and he promised to take me to a nearby reef where I would see some fish we wouldn't get in the bays. Soon after our arrival I had begun swimming in the area, close to shore, wearing a face mask, but I was not overcharmed by these semitropical, semiclear waters. I'd been spoiled by diving in the clear water around coral reefs in the Bahamas, the tropical Pacific, and the Red Sea. And there was much to do setting up the Lab, the new room, the pen, the work with sharks, and the classifying and preserving of all the material we collected in the bays from the stop-net fishermen and from days of seining along the beaches. Sometimes Hera, young John Bass, Bill Jr., his older sister Elsie, and her husband, and Alfred Vanderbilt's children Heidi and Butch joined us and helped pick out all the small fishes we caught in seaweed and in our fine-mesh seine. I hadn't thought much about diving. But the late-spring heat was getting strong, and it did seem like a good idea to dive and see the possibilities for study in slightly deeper water.

Beryl kept the *Dancer* anchored on the reef about three hundred yards directly off the Boca Grande Range Light on Gasparilla Island while I took my first dive away from shore into the Gulf of Mexico.

I found myself in calm, murky green water with no point of reference. The water was only 16 feet deep here, yet halfway down I felt like turning back, and as I paused I almost lost my sense of which way was up. Better keep going, I thought. What a short distance to look endless! I continued swimming down through a fog of rich plankton and nonliving suspended specks, in water laden with nutrients. Then, as if a magician's puff of smoke had cleared, a fairyland appeared: a reef of brilliant orange-red rocks broken here and there by growths of purple and green algae and blobs of plantlike invertebrates.

It was not a coral reef; no coral reef would have shown such an expanse of red. Actually there is little true coral in this area. I had to look hard to find an occasional patch of coral, and then it was no bigger than the palm of a baby's hand. The red color of these rocks is caused by an encrusting sponge, *Cliona*, which forms a soft sand-to-sand carpeting over the irregular limestone outcroppings. A type of *Cliona*, more orange, lives in nearby bays, sometimes giving the lining of oyster shells a case of "measles" and finally riddles the shell with so many borings that the oyster's hard covering crumbles away.

Sprinkled over parts of the red carpeting I found a more spongelike-looking sponge. The bright yellow *Verongia* varies from the size of pimples to elongated, hollow, branching vases over two feet high in which creeping clinid fishes hide and nest. In places, the sponge carpeting gives way to patches of algae. Some grow like green hedges regularly clipped; others form jungles of variegated shapes. A cluster of thin, yellow-green smooth hairs grew from one spot and I felt like combing it. From another spot a tangled mess of coarse strands emerged, the ends of each strand curled in a clutching grip as if they grew from the head of a buried Medusa. Rubbery fronds of rose-colored algae, which felt like Jell-O left too long in the refrigerator, slipped between my fingers. (Later I took some of this home and made a delicious custard, which I call beachcomber's custard: One handful of *Eucheuma* alga simmered 10 minutes in a quart of milk. Strain the milk, add sugar and vanilla flavoring, pour into custard dishes, let cool, and set in refrigerator until ready to serve.)

Colonial tunicates, looking more vegetable or mineral than the animals they are, grow abundant in this area—large amorphous blobs of orange, black, lavender, and gray. When torn loose by a storm and washed ashore, the gray ones are sometimes picked up by beachcombers who come to the Laboratory to ask if they've found ambergris.

A hundred fish were concentrated in the small area I could see. I looked around hurriedly at sizable food fish— snappers, grunts, groupers. Then I peeked under a ledge in the red rocks and, feeling as if I were looking into a doll house, saw miniatures in the shadows. A tiny sea bass two inches long looked up at me, then turned away clumsily. It was a fish I'd never seen before. It was weighted by a swollen belly that flashed conspicuously white compared with the dull reddish color of the rest of its body. This baby-sized grouper appeared pregnant with ripe eggs. It aroused my curiosity, but the casual first meeting with *Serranus subligarius* lasted only a few seconds. In the grouper's place came an undulating shivering, ribbon. A young high-hat fish, an exaggeration of the adult, was startled by my hand. With short uncertain spurts of movement, it retreated deeper under the ledge, trailing fluttering long fin attenuations from its back and tail as it turned from side to side, glancing back at me as it fled. It seemed incredible that this odd-looking creature should belong to the same family as the ordinary-looking weakfish and the spotted sea trout. Above the ledge, the red carpeting was honeycombed with marble-sized holes. Out of each opening, the face of a small blenny fish, with eye makeup like a mime, stared at me. I had never seen a reef like this and. I peered at the marvelous little creatures until I had to push off for air at the surface. I kept diving to see more of the bottom community, wasting much time and effort going up and down for breaths, and wishing I had scuba gear.

On every dive to the bottom that day, I saw something worth watching longer than my breath could hold. A blenny ventured out of its hole to catch a tiny shrimp and then tried to argue a neighboring blenny out of its shelter. The two locked jaws in combat and risked being the double hors d'oeuvre of a passing grouper. Triggerfish, surgeonfish, butterfly fish, and sheepshead, larger members of the red carpet community, swam close by. The "blue" tang is bright yellow when it's small enough to huddle under the shallow ledges with the other miniatures.

The tiny white-bellied grouper was the most common fish around. Pairs and trios lived under the ledges. Each one seemed to be a female with a swollen belly ripe with eggs. Where were the males? They should be around when females are ready to lay eggs, but I couldn't see one.

I wondered how I might catch some of these fish and study them in an aquarium at the Lab. On the way up from my last dive for that day, a large form passed close to me in that mid-water fog. It was just at the limit of my visibility—a fish about five feet long and with a deeper body than that of a barracuda but I couldn't make out what it was. I climbed into the boat quickly.

Beryl enlightened me. "Did you see the school of tarpon go by?"

"I caught a glimpse of just one," I answered, relieved that it was only a big cousin of the sardine. But I knew it would take a lot of experience in the Gulf before I got used to the new sensations of

The lab's first shark boat, named after the Vanderbilt's race horse Native Dancer, with our fish collecting skiff in tow.

*Cape Haze Marine Laboratory, Placida 1955
Our original building was 20'x 40'*

diving in water with such limited visibility.

The central Gulf coast of Florida is famous for its tarpon fishing. Sportsmen from all over the world come to this area, especially Boca Grande, where Hemingway enjoyed and wrote about the sport of catching "silver kings." Adult tarpon, which average over 100 pounds and may go to 300, put up a terrific fight on the end of your line and half an hour is considered good time to land one.

Beryl was an enthusiastic rod-and-reel fisherman and a great tarpon fisherman. He made the headlines, however, and greatest claim to tarpon-fishing news when he landed a medium-sized tarpon. He told me the story of this unique incident as we brought our boat hack to the Laboratory.

When Beryl was younger, and "didn't know better," as he put it, he tried an unorthodox method of catching a tarpon. He was standing waist high in water near his house on Lemon Bay when a school of tarpon swam slowly by. One came alongside, its head less than a foot from Beryl's right hand, and a wild impulse overtook Beryl. He quickly slipped his hand deep into the opening under the fish's left gill plate. His arm was almost pulled out of its socket as the startled fish took off. The commotion and splashing attracted a crowd of people who watched the struggle as man and fish flayed about in the shallow water. When Beryl finally dragged the tired tarpon to shore, blood was pouring out of the gill opening,

"My blood it was," he said. "My hand was caught in that fish and its gill rakers were hooked on to the flesh of my forearm. People were congratulating me for holding the tarpon till the end. But, hell, I couldn't have let go of that fish if I wanted to."

Beryl had spent over a quarter of a century catching big game fish as a hobby, and professionally he caught food fish by tons when commercial fishing was at its peak in Florida and his father and uncle, Chadwick Brothers, Inc., operated a fleet of fishing vessels in the area. Beryl had also been a charter-boat captain guiding anglers looking for big fish. I was fortunate that this experienced fisherman who handled specimens that sometimes weighed hundreds of pounds was as interested in tiny fish a fraction of an ounce.

When I described with excitement the little fish under the ledges, Beryl told me of some fish he used to find as a boy— fish that lived among clusters of oysters, in tin cans and bottles, and in the pile of discarded clamshells in the bay beside his home. He wanted, as much as I, to have a whole room for aquariums with running sea water and large fiberglass tanks, where we could keep and study many of these small fish. He had started already to construct the pipe lines and pumping system for our new lab room. I visualized aquariums filled with all kinds of marine life and anticipated the fun of collecting them with Beryl, studying the animals both in natural conditions where we could find them and under the more artificial but controlled conditions of a laboratory's aquarium room. I felt incredibly lucky with my first full-time job, doing what I always wanted to do most—study fish—with everything all in one place, collecting grounds, a laboratory, and my home and family, who liked our new life in Florida as much as I did.

When Beryl and I closed up the Laboratory at the end of each day, we went home to our families. My day of diving wasn't over then. My two-year-old daughter, Hera, was usually waiting for me on the front steps in her bathing suit, miniature frog feet and face mask ready in her hands. I often brought her home some presents in a pail of sea water: a nudibranch mollusk, with a royal-blue and yellow-striped mantle, whose edges undulated and head tentacles waved searchingly as it crawled across the bottom of the pail; an orange starfish with brown warts; and a small purple sea fan I had picked from the reefs. We would put them in a glass bowl in the kitchen and take Aya from her crib to show them to her. Then the three of us would go for a swim along the white sandy beach which was our back yard; Aya clung to my neck and Hera dog-paddled alongside. The sea and beach were empty of people, and seemed to be ours. There was no Jones Beach policeman to advise me to put Aya's diapers back on.

If Ilias came home from work early, he would join us quickly. This was a happy time of day when we could relax on the beach with our children. We exchanged accounts of our day, talked about the promising future now that we lived in Florida. Within a year, Ilias hoped to open his own office, near our home. Having joked about it at first, my parents really were moving their Japanese restaurant

My family. L-R Ilias, Tak, Aya, Hera, Niki

Chidori from New York to Englewood, where they had bought property to build the restaurant and their home in a development called Grove City, halfway between our house and the Lab. Chidori became the first Japanese restaurant in Florida and the local paper gave it a lot of press. No one expected it to be successful in a little fishermen's village but it became very popular.

Ilias was working with an orthopedic doctor, Mike di Cosola, in Sarasota, forty miles away. It was a long drive. In his spare time, he was studying and reviewing for the Florida State Medical Board exams in order to practice medicine on his own. Yet having a home on a beach in Florida, where we could go for a swim in the morning or evening and spend hours weekends walking and playing in the sand, after life in New York City, gave us the feeling of being on a continuous vacation. Hera and Aya were healthy, suntanned, and totally unafraid of the water, and they accepted all that went on in our lives so naturally. Hera's only jolt after her city life in New York was her utter amazement and surprise at the sight of the full moon. We realized when we showed it to her that she had never seen the moon before. It is a beautiful sight in Florida's clear sky, with palm trees and water to silhouette and reflect its bright light. Hera came crying to us a few nights later to inform us that someone had "bwoke" the moon.

In the evenings after we put the children to bed, Ilias settled at a desk and studied in his fashion, underlining in his books important words and phrases,

With my mother

30

never taking notes. I liked to sit curled at one end of the sofa with books and articles on an end table and spread on the floor, scribbling notes and sketches on a large pad on my lap. The breathing of two young sleepers and a soft surf were the regular sounds close to my ears during the quiet hours before midnight, The peace was broken only occasionally by the metal garbage can being knocked over by "Uncle Yoshi," a raccoon who often paid a visit around that time of

My friend Teru Kurokawa and I as teen-agers in New York studied dance together.

night. Those hours gave me time to wonder about the problems and mysteries of sea life that were part of the studies developing at our little marine laboratory in Placida. After my dive in the Gulf on the red reef, I particularly wondered about the males who should have been around—for all those little female groupers ready to lay their eggs—and weren't.

2 The Lab's First Year

IN A SCIENTIFIC REPORT, one doesn't confess that a study was started because was "I was looking for an excuse to go swimming." But in spite of the highfalutin reasons we marine scientists sometimes give for our projects, many projects have started with such a simple initial motive.

As the summer of 1955 progressed, I began thinking of a project where I would spend less time in the hot Lab rooms and more time in the water. In July, I attended the annual meetings of the American Society of Ichthyologists and Herpetologists in San Francisco, where I heard an inspiring lecture given by Connie Limbaugh, the Chief Diver at Scripps Institute of Oceanography. He told about his method of catching small bottom fishes in a glass jar.

I decided to try to make a census study of the fish population on the colorful reef near Boca Grande Pass, and to use the glass jar technique to catch specimens I couldn't identify. A record of the species of fishes inhabiting a reef and their relative abundance, estimated in some quantitative way, can give marine biologists an idea of the fish biomass supported by such a reef.

That summer and fall, I spent a good deal of time exploring the reef and recording its inhabitants. The glass jar technique worked but it took awhile to learn how to do it. In water 16 feet deep, I caught a fish on the average, I guess, of every six dives. Baby ribbonfish, young surgeonfish, blennies, gobies were all caught in a half-pint mayonnaise jar. The prize fish, however, was the little grouper with the swollen belly I had noticed on my first dive.

Each type of fish requires a slightly different approach and some understanding about the habits of the individual fish one goes after. Each little grouper, for example, had a definable territory. On the first dive, I'd make a reconnaissance. I'd try to find a good ledge with at least one of these tiny groupers under it. I'd make a mental map of the area, lining up the ledge with some landmark such as a tall yellow sponge. Surface for air. Then down again, hoping to come out of the fog at the same spot below, but even with luck it took a few seconds to locate the same spot. (Occasionally I never found the spot again!) Then I'd have about fifteen seconds to chase the fish with a stick and study the way it moved about its territory and where its favorite hiding places were.

There was usually a definite pattern. If I approached a fish under a long ledge and put the glass plate of my face mask fairly close to it, the fish would scurry along the ledge close to the back wall. Its direction to the left or right could be controlled by prodding it. Chopsticks from my stepfather's restaurant made good prods. A flat stick slightly smaller than a ruler, with a sinker strung through one end and tapered on the other, makes an even better, more sophisticated prodder. A hand works best if it doesn't cause its owner any trouble when a stone crab, a toadfish, or an ocellated moray, which might also be under the ledge, snaps at it. A face mask just outside the ledges is rarely attacked by these animals with formidable claws or teeth. The only fish that had the courage to attack the face mask I was wearing was a two-inch-long gold-and-blue damselfish who was guarding her eggs.

If the diver slides his face along a close parallel, as the fish moves, the diver's face mask, acting as a magnifying lens, gives a good view of how the fish follows the irregularities of the ledge, when it turns a corner or stops and quivers at the end of the demi-tunnel. You can sometimes prod the fish out of the protection under the ledge and along gullies in the red carpet, where it may then dart into a deadend hole, and you can gently poke it out with a chopstick and trace its next course, often back to the same ledge again in some roundabout pathway, always staying close to the substrate. If you chase the fish too much, you can drive it out of its territory and then its movements are too fast and unpredictable to plan a trap. It will swim down any hiding hole and often come fleeing out chased by a prior resident who owns the hole. Surprisingly, big stone crabs put up with these little intruders best.

Once you've mapped the territory and gently nudged the fish around to learn its pathways (maybe on the third or fourth dive), you can plant your jar with success. Just around a corner is a good spot. Lay the jar on its side, mouth in the direction the fish will come, and don't let the fish see you do this. Then you go back to the fish and prod it in the direction of the jar. The fish invariably seems to take longer this time, probably sensing something is about to happen. You don't hurry him or he might flee his territory. You want him to stay in his regular channels. Then comes the crucial moment. He approaches the turn and you follow him slowly, your face mask within inches of the nervous fish about to dart around the corner into your jar. Alas, this is usually the moment your breath absolutely gives out and you go to the surface, gasping as your fish darts into the jar, bangs his nose on the glass bottom, and turns about and exits like lightning.

But if your breath holds out, and your hand can follow behind the fish and close over the mouth of the jar as he enters, you've got your specimen. Otherwise, you start from scratch again, and better try another fish.

Sometimes other things go wrong. Once, as the crucial moment arrived, I saw out of the corner of my eye a crab come sauntering along the ledge from the opposite direction. His claws touched the jar, he felt around a bit, and then with a gentle push he rolled it away and the fish turned the corner, swam past the crab, and got away.

I learned to wedge a stone or shell under the jar, for it rolled easily, sometimes just from currents on rough days. It was frustrating at times but always fun and a challenging game to try to outwit the fish. Specimens caught this way were unharmed and lived well in the Lab's aquarium room, especially the hardy little grouper,

As I became familiar with this reef and its inhabitants, the tiny grouper with its white swollen belly became an ever greater puzzle. I dissected several. Some had full stomachs with a freshly swallowed blenny or a snapping shrimp. But the main distention of the belly in every case was due to an enlarged ovary full of ovulated eggs, broken loose from their follicles in the inner wall, filling the central cavity of the ovary,* and ready to be released and fertilized—the typical condition in ripe female bony fishes whose eggs are fertilized externally.

Again and again came that puzzling question that occurred to me on my first dive: Where were the males? After inspecting the reef many times, I learned to identify every fish I could find. There was no possible mate, even in the disguise of a different sex coloration or form. No mate! No source of sperm for the hundreds of unfertilized ovulated eggs inside each of the thousands of *Serranus* living in that reef? Every *Serranus* I saw, from two-inch specimens to the largest, about four inches, was swollen, and with eggs I was sure. Those eggs had to be laid and fertilized within a short time or go to waste.

The fish we brought back and kept in aquariums seemed gradually to flatten out in the course of two or three days.

We kept them isolated in separate tanks, as they seemed to fight when we placed two together. We fed them all they could eat. They ate well but never became swollen any more.

Sylvia Earle, an attractive young marine biologist, was becoming a regular visitor at the Lab. She was studying algae for a doctorate degree at Duke University and often came to see her parents, who

---

* The teleost (bony fish) "ovary" is really a specialized part of the oviduct, and therefore has a central cavity or lumen. The ovary of higher vertebrates and cartilaginous fishes (sharks and rays) is solid, and the eggs are shed from follicles on the surface into the abdominal cavity and then swept by cilia into the tubes leading to the uterus.

lived in Dunedin, Florida, and then came to the Lab. During her first visit, she offered to help me start a reference collection of marine plants. Beryl and I had great respect for Sylvia's amazing ability to untangle the masses of what Beryl and the stop-net fishermen called "gumbo" and to call off the scientific names of twenty kinds of algae, modestly telling us she wasn't sure of every name but her professor at Duke, Dr. Harold Humm, would check her identifications before she labeled and deposited her duplicates in the Lab's reference collection.

Sylvia loved to dive as much as I did, and I suspect that for her, too, the chance to work in the water was a strong motive in planning her research projects. I was particularly happy to have, on the red reef off the Boca Grande range light, a diving companion who appreciated the marvelous variety of algae and other biota growing there.

In September, Beryl took Sylvia and me and other divers on several trips to collect samples of every kind of algae we could find on the red reef. On one of these trips, I met a manta ray underwater while diving for algae and trapping little fishes in my glass jar. The water was relatively clear that day. I felt the water above me stir before I looked up and saw the giant fish. I could make out immediately that this moving white roof was the underside of the largest species of ray in the sea. The manta looked me over as it swam gracefully around me. Other divers in the water and people in the boat also got a good view of the manta. It stayed around us for nearly an hour, often with one or both of its wing tips protruding and cutting the surface of the water like the dorsal fin of a shark or pair of sharks.

We saw several other small mantas in the Gulf that week and

decided to try to catch one. These plankton-feeding rays had never been kept alive in captivity. Their food was a problem. Beryl planned to use our fine-mesh seine to get batches of the tiny sardines and anchovies that formed massive dark clouds in the water and were easy to collect in quantity. Then he was going to ask his wife to put the planktonic fish into gelatin balls he could roll into the manta's mouth once we got one in the pen.

Beryl went to Johnny Bass and borrowed his Swedish harpoon gun. Beryl got the smallest manta we could find on the first shot. But we kept it alive in our pen for only a few days. The harpoon had penetrated the liver and other viscera, and it bled a lot. It was an immature female that was 7 feet 8 inches wide and weighed 257 pounds. I had a few years to go before I had an encounter with a full-grown adult.

Our trips to the red reef were highlights of the two or three days of the Lab's weekly schedule we allowed for field trips. We spent mornings diving and then took the boat south through Boca Grande Pass to the town of Boca Grande and docked on the bay side of Gasparilla Island. We had thick hamburgers for lunch at the Pink Elephant. And Beryl refueled the *Dancer* at the nearby public dock.

It was a longer trip back through Gasparilla Sound than straight along the Gulf side but we took this route not only for variety and a different type of collecting but also because in the afternoons the wind usually came up and the Gulf became rough. Beryl took us through the narrow bayous where we could see many kinds of water birds perched on the mangrove trees, the nest of a bald eagle, and an island dense with nesting herons, egrets, cormorants, and pelicans. Oysters covered the mangrove roots exposed at low tide.

We'd stop, turn off the motor, and eat some oysters. The peaceful silence was broken only by the splashing sound of leaping mullet or a bird's cry.

Sometimes the tide was so low that the tops of the long leaves of acres of grass beds would be touching the water's surface. If we took time out to stop and wade through one of the beds with nets, we could collect shrimps, crabs, sea cucumbers, and thousands of baby fishes in this nursery and feeding ground for many of Florida's most important food and game fish. We also liked to anchor next to an exposed mud flat and walk across it, digging into the mud volcanoes with a shovel and picking out worms with head tentacles more elaborate than ladies' Easter hats and locomotor organs along their bodies like feather boas.

Large pen, olive, and angel wing shells with the animals inside also came up with a shovel's turn, and once while I was doing this my foot sank into the mud and I felt a tiny painful bite. I pulled out my foot and shoveled up the mud. I found only a smooth pink worm six inches long, but when I put it into a bucket of sea water I saw the worm suddenly throw out a club-shaped proboscis, a hidden head with four black fangs. It took a week for my swollen and itching foot to heal from the bite of *Glycera*. Beryl never got bitten. Even when, dozens of times, he'd stick his hand into depressions in the mud flats, feel around blindly for a few seconds, and then each time come up with one or two huge stone crabs.

When we reached the railroad that was the only direct connection between the mainland and Big Gasparilla Island in those days, we'd have to blow a horn for a man to come out of his cubicle and hand-wind the railroad bridge open. Then it was only a short distance to the Lab.

But one afternoon the Gulf was still calm and the water exceptionally clear, and though it was getting late we all wanted to go back for a few more dives at the red reef. I was in for a surprise. I went from ledge to ledge scrutinizing every *Serranus* I could find. Every one was flat-bellied! They had laid their eggs with no males around! What happened between the hours of 1 and 5 P.M.? The Lab had to invest in scuba gear for me to find out more about the activities on this reef. But it was near the end of the *Serranus* egg-producing season. Before I got the scuba gear, they had stopped producing eggs for that year.

The first year of the Lab went by quickly. By the end of 1955, we had four well-defined rooms. The original reference-collection room lined with shelves was filling up with preserved fishes and invertebrates that Beryl and I collected, catalogued, labeled, and arranged in groups according to the families in which each species belonged. Here we also kept our marine herbarium, folders of pressed and dried algae that Sylvia Earle and Dr. Humm helped us with. Near the center of this room, across from the sink and side tables, a large central table held our microscopes.

An aquarium room 12 by 20 feet (slightly larger than the collection room) had over 30 aquariums, varying in capacity from a few gallons to 20 gallons, plus a series of larger porcelain holding tanks for sorting live collections and keeping some of the larger animals. Beryl had made the latter out of the liners of old refrigerators. A pump kept them supplied with running sea water. As our needs for keeping animals in sea water increased, we gradually cut down on our cages and boxes with "dry" animals, until we kept only one cage with a five-foot corn snake and a small wild mouse Beryl had caught in the field. The mouse had been put in as food for the snake, but just after its introduction, as the snake raised its head and was poised to strike at its intended victim, the mouse jumped up and bit the snake's nose. The snake left the mouse alone after that, and Beryl began sharing bits of his lunch, like a slice of orange, with it. The mouse sometimes rode on the snake's back as it crawled around the cage, the mouse standing on its hind limbs while holding and nibbling on orange seed. This strange association ended only when Beryl decided the mouse deserved its freedom and let it loose in the field again.

We also had a small office and file room and a sizable library room. We were accumulating all the books we could find and fit in our budget pertaining to marine life on the Gulf coast of Florida. We started subscriptions to scientific journals and much of my paper-work time was devoted to writing people and organizations to get the Lab's name on the mailing list for scientific papers on marine life. The Lab had become incorporated as a nonprofit organization, and we had to keep books and account for almost every penny we spent. Neither Beryl nor I knew much about bookkeeping. Fortunately, Bill Vanderbilt shared one of his secretaries, Marion Suss, with me. Marion was the Lab's secretary, bookkeeper, librarian, and in a pinch helped measure and photograph sharks, clean and dry fishbones. In contrast to the way Beryl and I dressed (suitable for wading into the bay on a moment's notice), Marion came for her half day's work in dress and neatly bobbed hair appropriate for the offices of the Cape Haze Development Corporation. She was a highly efficient secretary and she quickly mastered the most difficult scientific terms. Marion came from New England. Her trim, almost prudish appearance was deceptive. One day after I'd dictated some letters to colleagues with whom I'd worked in

Egypt, Europe, and some far-off spots in the Pacific, telling them of the Lab's work and requesting reprints of publications, Marion let down her usual reserve and commented that she found the work fascinating and even the correspondence exciting. I thought it probably was an unusual experience for an office secretary. It was not until later I learned that Marion was the wife of Herman the Great, a retired famous magician-illusionist who had created some of the most extraordinary acts involving mirrors, electricity, and a glamorous female assistant who entered his mysterious boxes. And Marion had been his assistant!

During the Lab's first summer, we had a variety of children and young students visit us during the day. Many asked to come back again, and one stayed with us for a solid month. Carey Winfrey was the son of the man who trained Native Dancer, among other horses, for Alfred Vanderbilt. Carey, who was only thirteen years old, convinced his father he could live alone in a motel near the Lab and keep himself busy and out of trouble. He had a mind for taking things apart, keeping track of minutia, cleaning and putting things together again. He could easily have been a watchmaker. In the evenings, he worked on a radio he was building from parts. At the Lab, he spent much time boiling up the heads of fishes, cleaning the dozens of bones that fell apart in these very complex skulls, then gluing them back together again. Sometimes during lunch hour, Marion Suss would come over with her cheese and crackers and thermos of cold drink and sit beside Carey, looking over his shoulder and suggesting where a bone might go as he put together his jigsaw puzzle. Carey dived with me at the red reef off Boca Grande, and we shared a case of poisoning.

Dr. Lawrence Penner, a parasitologist from the University

of Connecticut, brought his wife and six children and two lab assistants to work with him for the summer. We had to take the Penner tribe in shifts on field trips in the *Dancer*. Carey, several Penners, and I encountered a "bloom" of thimble jellyfish* on the red reef. The jellyfish, the size and shape of thimbles, were so thick in the water that we were touched by hundreds, perhaps thousands, in the two hours we spent diving on the reef. None of us was sure the jellyfish would sting us, and before we dived we scooped some up in a bucket and put them on our skin. We couldn't feel any sting, but as I dived I thought I felt slight pricks, especially when one brushed across my lips. The next day, Carey and I came to the Lab looking like albino Ubangis with measles. The jellyfish poisoning developed during the night. But the Penners, the healthiest family I've ever encountered, were completely immune.

By the end of summer, we had two other teen-age volunteer assistants. Bruce Marshall and Bill Brown helped Beryl, Marion, and me and visiting scientists with all kinds of odd jobs.

By the end of the year, twenty-eight scientists had come to visit and work at the Lab for a few days or sometimes about a week. Many discussed with me plans to come back for a longer visit the following year with research projects. Beryl and I could advise most of them when the best time of the year was to collect and study the particular marine organisms they were interested in. I looked forward to visiting scientists; each one had different needs, and as Beryl and I tried to help them we learned many things ourselves. One scientist, Dr. Edmund Güttes, requested a particular alga, *Acetabularia*. He was not an algologist but a cell physiologist experimenting on how cells function without a nucleus, and *Acetabularia*, which looks like tiny green parasols, each half an inch high, had giant cells. Each parasol was a single cell, with the nucleus at the bottom of the handle, which could easily be cut off from the rest of the cell.

My family continued to take pleasure in Florida living. Hera and Aya enjoyed all the extras of having doting grandparents around, and I knew I could go on with work at the Lab, even with a third child on the way, since my mother was always available and pleased to help with baby-sitting.

Sometimes my mother came to visit the Lab with Hera and Aya. Beryl would let her use his fishing pole and show her his favorite fishing spots. She'd spray my daughters and herself with mosquito repellent and stick it out until she had our supper: a black grouper

Sylvia Earle, collecting algae for her doctoral research

Beryl Chadwick, the Lab's first shark fisherman supreme.

Students collecting fiddler crabs. Note the lab addition on right.

or some mangrove snappers. I don't particularly like rod-and-reel fishing—waiting for a fish to bite. I'd rather go after fish with nets, a glass jar, or a spear. But I like to clean fish and look at their insides in detail before cooking them. Hera and Aya were always an attentive audience whether I opened up a large shark at the Lab or a grouper in our kitchen. Young children are fascinated by the innards of a freshly opened animal—the red blood, the yellow eggs, the shiny purple liver, the pink pancreas, the slippery coils of intestine. Children's first anatomy lessons should be in the kitchen while their mother cleans a fish or chicken. Packaged supermarket foods reduce this possibility, but fathers will probably never give up sport fishing, and bringing home their catch.

Uncle Yoshi, the raccoon, continued to visit us in the middle of the night. We had named him after an old friend, Dr. Yoshi Kondo, the malacologist at the Bishop Museum in Hawaii, whom I first met when I studied fishes in the Pacific. Just before we moved to Florida, Yoshi was studying snails for several weeks at the American Museum of Natural History in New York, and he lived in a small room nearby. He refused to stay with us, as he explained that "fish and house guests stink after three days." Yoshi kept an odd schedule. He was a fine scientist, and his publications revealed the intricate and detailed anatomical studies he made on mollusks. But no one ever saw him work much between nine and five o'clock. He always seemed to be occupied with visitors and helping other people with their problems, or discussing life in general and his own unique philosophies. In New York I discovered how he managed to do this. When I invited him for supper, he liked to help me cook or to take over the cooking completely, usually bringing a fish that he'd slice up for sashimi. He didn't want to get into any deep discussions but preferred to play with Hera. He usually dozed off before dessert and was snoring loudly by Hera's bedtime. Hera, Ilias, and I would retire for the night and leave Uncle Yoshi on the couch or in an armchair. No matter how early we got up in the morning, Yoshi was gone and we'd find that he had cooked a fresh pot of rice before he left. He finally confessed that he loved to eat fresh rice in the middle of the night, so I gave him a key and told him to feel free to cook rice in our kitchen whenever he wished. He did, regularly. When I came home from the hospital with Aya, I found out at what time. I woke up at 2 A.M. to give Aya her bottle, and Yoshi was just finishing his third or fourth bowl of rice and starting off the important part of his day; he was ready to go to the museum well rested, his belly full of his favorite food, and do his real work in the wee hours, free from any interruptions. The second night I woke to feed Aya, Yoshi had already warmed her bottle and was feeding her, "Go back to sleep, imotosan," he told me—and he took over Aya's 2 A.M. feedings.

Uncle Yoshi, our raccoon visitor in Florida, also liked rice. He gradually lost his timidity and came around when he smelled our supper cooking. If we left the kitchen screen door ajar, and didn't look directly at him, he would join us in the kitchen and Hera and Aya shared their supper with him. He continued to raid our garbage can in the middle of the night and grew quite fat. Then one day he didn't show up. We began to worry by the third day of his absence. On the fourth day, "he" reappeared, quite slim, with three baby raccoons.

One event made a dark mark on our first year in Florida. In November, Ilias's car slid off the road and crashed into a tree on his way home from work. No bones were broken but he was knocked unconscious. My mother took care of Hera and Aya, and Beryl took care of the Lab for the three weeks I stayed in Sarasota near Ilias while he slowly regained consciousness in Memorial Hospital. By the end of the year, his memory had still not wholly returned, but he was recovering slowly and in time was to continue his practice and perform operations in the same hospital he had been released from as a patient.

3

Sharks and Abdominal Pores

IT WAS WHEN I was at Hunter taking a sophomore course in the comparative anatomy of vertebrates that I first saw abdominal pores. They were in a dead and preserved dogfish, a small shark used routinely for study as a typical example of a fish. After studying aquarium fishes, I found that a shark didn't seem to be very "typical," and therefore it was especially interesting.

The study of the dogfish—or any preserved animal, for that matter—usually begins with an examination of its external anatomy. What kind of skin and limbs has the animal? Which is the head or cephalic end, which side is dorsal (back), ventral (belly), left, right, up, and down? How many external openings are there? These elementary questions sound simple when you consider a man, cat, canary, or cockroach, but it becomes a guessing game when you ponder over a proboscis worm, an upside-down catfish, a sea urchin, a snail, a sponge, or a jellyfish. Or try to answer just one of those questions when you look at an amoeba—which you must do under a microscope. The amoeba has a cell membrane with no opening, yet any part of the membrane can admit or reject objects in order to eat or excrete.

The body openings of higher animals are relatively easy to define. If one eliminates such microscopic pores as those involved in goose bumps, perspiration, and other surface excretions, man—once he is born and his umbilicus closes—has nine body openings. A woman has ten. The eyes, ears, nostrils, mouth, urogenital, and anal openings are similar in all mammals. As we move down the evolutionary tree, the birds and reptiles have the openings on the head distinct, but the other openings in the anal region are sunk in a pocket, the cloaca, which neatly closes and streamlines the cloacal region. Farther down the evolutionary tree of vertebrates, the young amphibians and fishes have a cloaca and as many additional openings as they have gills, and sometimes a few more. Sharks living today may have as many as seven pairs of gill openings, although usually the number is five, with occasionally another pair of small openings, the spiracles, which are rudimentary gill openings that disappear entirely in some sharks and develop into fascinating large breathing organs in rays and skates, relatives of the shark. You can pick up a good-sized sting ray by slipping your fingers into its spiracles. It's like picking up a bowling ball and is the best way to lift this slippery creature and avoid the poisonous sting on the tail. Just don't pick up an electric ray the same way. The shocking organs are next to the spiracles.

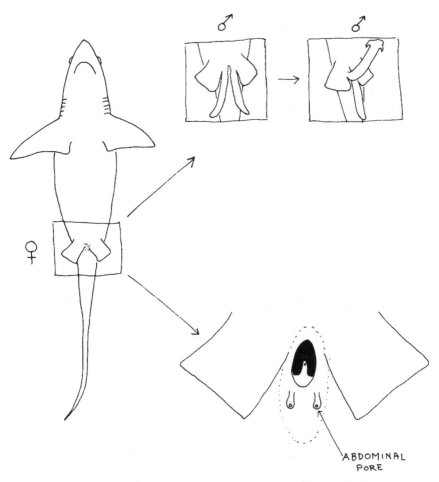

ABDOMINAL PORE

The nostrils of fish and other non-air-breathing vertebrates are usually dead-end depressions for the sense of smell only, rather than openings for a breathing passage, as in man. And in water, where sound travels faster and louder than in air, external ears to catch sound are needed less, and fish lack these paired openings too. But primitive fish have a pair of openings all higher vertebrates lack—the abdominal pores.

To find the abdominal pores of a shark, you must spread open the cloacal region. This you do in a comparative anatomy course when you probe into every nook and cranny of your animal. The pores are easy to find. Each is located at the tip of a pair of small teat-like structures called abdominal papillae. The pores open directly from the main body cavity (abdominal or peritoneal cavity)—the cavity that in man so neatly holds all the viscera in closed circuits. Stomach, intestines, reproductive organs, etc., may expand in your abdominal cavity when you are pregnant, gassy, or both. But ordinarily nothing from these systems leaks into the abdominal cavity. When it does, you are in trouble. Any invasion of this space by a bacteria-contaminated fluid or material promptly results in a most painful and dangerous-to-life inflammatory reaction (acute peritonitis). The other complication in man involving the peritoneal cavity is called dropsy or ascites, which refers to a rather slow accumulation of sometimes enormous amounts of fluid in the space.

A typical shark doesn't need a drain for this cavity. It has two—the abdominal pores. Why should a shark have these openings connecting its body cavity directly with sea water? After I finished my day's dissection on the dogfish, I looked up the function of the pores, first in the class bible, Libbie Hyman's *Comparative Anatomy of the Vertebrates*, then in other books. One didn't ask the teacher, especially Mrs. Little, for answers one could find in a book. So I searched and found descriptions of the anatomy of the pores and associated structures. They have been well known for over a hundred years and had been studied by many scientists in many sharks—like mine, dead. I couldn't find the answer in books. At the next class, I asked Mrs. Little what function the pores served. She didn't give me a stern look of reproval for asking a question I could look up myself. She stammered a bit and told me that the pores were more functional in cyclostomes (vertebrates such as lampreys, more primitive than sharks), which have external fertilization and pass their eggs out of their abdominal pores. She told me fishes other than sharks had these pores, too, vestigially. They may be vestigial in sharks, too. Some sharks don't have them. But what they may be doing in live bearing sharks with internal fertilization, she didn't say.

"I'll speak to you about them in the next class, she said, and turned the back of her red hair to me as she hurried to get on with the busy schedule of her day. Oh, she doesn't know, and is going to look it up, I thought with the secret delight of a cocky young student who had caught the professor—a professor whose brilliant teaching I didn't really appreciate until years later.

At the next class, Mrs. Little told me simply that the function of abdominal pores in sharks is not known. How ridiculous, I thought. Why doesn't someone find out? When I get my hands on a live shark, I'm going to tie those pores closed and see what happens to the shark.

The first experiment I tried on abdominal pores was a complete flop. Dr. Heller could not get back as soon as he planned, but he kept in constant touch with me and was pleased to know that our large fish pen, plus a slightly smaller one that Beryl had added to the other side of our dock, was working out well as a holding place for live sharks. As soon as Dr. Heller could take time off from his work, he was planning to come back. His growing medical institute, which had started in a barn, was developing into a series of research laboratories with equipment that was elaborate and expensive compared to that of our relatively simple and small field station.

In the meantime, we got some three-foot-long lemon sharks from the stop-net fishermen, John Fulton and his sons, who worked the bay next to the Lab. Beryl held a shark quiet while I tied the pores closed by pinching out the skin around them and pulling a loop of surgical thread tightly around each pore. The shark swam around for four days and didn't seem the least disturbed by what I thought were closed pores. When I examined the shark, I found it had huge gaping holes in place of the pores I'd tied closed. I had cut off the circulation and the tissue had sloughed off.

I made artificial abdominal pores in other fishes—gobies, groupers, snappers—by putting small plastic tubing in cuts I made into the fish's abdominal wall. The fish could not control these artificial openings as sharks evidently can, so sea water could flow in and out of the fish's body easily.

They lived for months in this condition, which didn't seem to bother them at all.

In some other experiments, I injected dyes into the abdominal cavity of lemon sharks, then let them swim in a large plywood box lined with fiberglass that Beryl built. I watched carefully but could not see any dye come out of the pores.

A professor from Cornell, Dr. Perry Gilbert, who taught comparative anatomy and was active in shark research, came to visit the Cape Haze Marine Laboratory. He and F. G. Wood at Florida's Marineland had worked out an effective way to anesthetize sharks with a compound called MS-222. A solution sprayed into the mouth and over the gills of a shark makes the fish unconscious in ten to twenty seconds. Dr. Gilbert demonstrated this method for us on small lemon sharks. We then used it for longer operations on sharks when we had to keep them unconscious for ten minutes or more. When Dr. Heller came to the Lab shortly after that, he showed me how to make some purse sutures around the pores of an anesthetized shark in such a way that the pores were closed but the tissue didn't necrotize and slough off,

We still couldn't find any function for the pores, but with the anesthetic we had a means of handling large sharks, spraying the MS-222 down their throats when we first pulled them up on the line. We could then operate on them at sea or bring them back without a struggle. Once the MS-222 was washed off their gills with sea water, they recovered with no apparent ill effects.

I made careful dissections of the pores and associated structures and studied fresh tissues under a dissecting scope and preserved tissue slices under high-powered microscopes. Dr. John Bracken, pathologist at Sarasota Memorial Hospital, made histological preparations of the tissues surrounding the pores. Dr. Dorothy Saunders, a microbiologist at the Lab, worked with John and me interpreting the histological findings.

The openings of the pores are controlled mainly by erectile tissue, but at the base of the papilla on which each pore is located there are sphincter muscles which further control passage of fluids through the pores. The tissue of the papillae also contains nerve endings resembling pacinian corpuscles (which in mammals are related to pressure and touch sensitivity) and, for some strange reason, the encysted larval stage of many parasitic cestode worms.

In studying the development of the pores in our local sharks and reviewing all the literature I could find about these pores in other types of sharks, I found a correlation between the development of the pores and the swimming habits of the shark. If a shark stays on the bottom even in deep water (such as the sluggish nurse shark, the squat Squatina, the angel shark with large flattened pectoral fins resembling wings, or the black abyssal horned shark), then the pores are very small or nonexistent and the papillae poorly developed. If the shark is a species which swims around at different levels in the ocean, going from one depth to another, then the pores and associated structures are highly developed. In a large tiger, dusky, or lemon shark, a probe almost the diameter of a pencil could be pushed through the pore opening.

Newborn baby sharks of species with large pores have their pores usually closed. If they are born and kept in captivity in a shallow pool, the pores will stay closed. But if these baby sharks are put in a pressure chamber (I had one built for small sharks which looks like a flying saucer from outer space) and under conditions simulating a sudden dive to 300 feet and return, the pores will sometimes break open. A large tiger shark that lived in our shallow-water pens for several months had pores that apparently closed during this time.

Sharks don't have swim bladders, as do most free-swimming fish, and appear to use these pores in making adjustments through various pressure changes. But because of certain complications (gassy stomach for one, I think) I could not get consistent results in my experiments in the pressure chamber—using dyes injected into the abdominal cavity and sophisticated pressure-reading instruments some electronically talented friends designed for my use with sharks.

Fourteen years and several thousand sharks later, after some marvelously rewarding experiences working with live captive sharks, I have yet to find the exact function of the abdominal pores. But it often happens in basic research that while trying to find the answer to one question you discover something else. We learned things about sharks that surprised and stirred us to study them more.

Dr. Perry Gilbert (L) on a visit to our Siesta Key facility with
Col. Rodney Carmichael. Behind them is the experimental pressure tank
used to simulate deep water in the study of the function of abdominal
pores of sharks.

4 The Mystery of
Serranus

THE MONTH OF MAY, 1956, dragged for my family. Themistokles
Alexander Konstantinu was due to arrive early in the month, but
didn't make his appearance until the 25th. Like our girls, he was
"well done," with dark brown eyes and the look of a baby several
weeks old when he was born. His father refused to have the name
Themistokles shortened to Timmy or anything else people tried to
talk us into. Marion Suss had the best suggestion. "Why not call him
by his initials?" The name Tak has stuck in this country. His relatives
in Greece still find it easier to call him Themistokles, especially
after waiting through three pregnancies for the first grandson who,
as custom has it, would take the name of his paternal grandfather.

Themistokles' Japanese grandparents on my side were glad to settle for Tak.

Tak was a few weeks old when I tried out the Lab's new scuba gear. I had no trouble finding the *Serranus*, and they were full of eggs again. Dr. Charles M. Breder, Jr., came to visit the Lab. He had been my ichthyology teacher at New York University. He was now Curator of Fishes at the American Museum of Natural History. He took this position after the New York Aquarium at Battery Park, where he had been director, closed down. I told him about the *Serranus* puzzle, and he suggested that these fish might be functional hermaphrodites. I checked my sketches of the fish's anatomy and found that I had labeled a thin winding white piece of tissue on the ventral surface of the ovary "fat?"

"That could be the testis," Dr. Breder told me.

I knew that one sure method of finding out if it was fat or testes would be to make stained sections and examine the tissue under a microscope. In order to do this, a sample of tissue is frozen or preserved in formalin, embedded in paraffin, then cut into thin sections with a microtome, a knife blade similar to the baloney slicer in a delicatessen, but more delicate. The sections are then mounted on slides and go through a long series of baths to dissolve the paraffin and stain the tissue, so that different parts of the cell become different colors. Then a histologist can tell if the tissue is part of an ovary, testis, spleen, nerve, fat, etc., whether it is normal or diseased.

When I took a course in histological technique as a graduate at N.Y.U., Professor Harry Charipper's lectures were spiced with anecdotes that kept his audience awake even on the hottest days of the year. The lab work was not as appealing. The end result, studying the finished colored sections under the microscope, was a psychedelic detective game. But changing the slides through all the baths in the long series of Coplin jars was a drag. And I made the mistake of taking this course in the summer, when lab sessions (usually stretched over eight months during the academic year) were crammed into eight weeks. Our lab wasn't air-conditioned, and I can remember dozing off when I should have been moving a slide from one jar to another. The worst part, however, was studying blood cells. Blood slides don't need many changes, but to get that drop of blood on the slide we had to use our own blood.

Professor Charipper recommended, with a leering smile, that for the intensive summer course, when we needed blood day after day, we pick a spot near the cuticle of a finger and prick it repeatedly. Then a handy scab would form which could just be pulled partway off each day for a good supply of drops of blood without the need of jabbing ourselves over and over. We preferred to keep jabbing. The "sissies" had someone else do the job. By the end of the course, all my fingertips were sore from repeated pricks with no days in between to allow for healing. When I couldn't find an unbruised spot left on any fingertip, I submitted to the earlobe prick. The earlobe is supposed to be relatively insensitive, but I decided that if I ever have my ears pierced it will be under general anesthesia. I went back to my fingers.

When Professor Charipper's lab instructor Al Stenger announced to the horrified, weakened, anemic class "Now we're going to need larger samples of blood for other tests," we all felt like dropping the course. But then he said, "Come and get it, vampires," as he set out a beaker of animal blood.

So I'd been through it—the whole course in histological

technique. I was rusty, but I could still set up all those Coplin jars, and could tell from the finished slide if the tissue on the ovary of *Serranus* was testicular. But Dr. Breder had a more practical method to suggest. "Next time you open a freshly dead *Serranus*, pinch off a bit of the suspicious tissue, put a drop of sea water on it, and look at it under the microscope." I did. The slide was alive with a wriggling mass of spermatozoa.

Hermaphroditism is known in both plants and animals. Most plants are monoecious, have both sexes in one plant. Among the flowers, it is more common to find the pistil and the stamen on the same flower than to have one flower a male and another a female. Among animals, only some invertebrates were known to be functional hermaphrodites, animals able to fertilize themselves and produce young.

Hermaphroditism among vertebrates is known as a freakish condition (old-time circuses exhibited such human freaks), but these people or animals are not functional hermaphrodites. Many are sterile and not capable of functioning as even one sex, let alone two, as the legend claims the offspring of the mythical Hermes and Aphrodite could do.

Among the vertebrates, certain fish species are known to be hermaphrodites of the protandrous (first a male) or protogynous (first a female) varieties. In these the individual starts off as one sex and turns into the other, functioning in one lifetime as both, but not at the same time.

Some of these fishes had been examined during the period of sex change-over when part of the organs of one sex still remained and the organs of the second sex had also developed. Such an individual might be capable of self-fertilization, but it had never been proved and is unlikely, as the sex change appears to take place during the non-breeding season.

No one had ever studied a vertebrate where every individual of the species could, at maturity, function simultaneously as both a male and a female and be able to fertilize itself. But this is the kind of animal that *Serranus* turned out to be.

I was amazed the day I first watched one of these fish isolated in an aquarium, resting on the bottom and releasing eggs. The eggs, each of which contains an oil droplet, float to the surface. I could see the eggs coming out of the fish and rising to the water surface. I scooped them out with a fine net and examined them under the microscope. They had just been fertilized. The sperm that penetrated the egg membrane left a scar at the entrance point and the fertilization membrane had started to lift on the egg's surface—a sign that the male and female elements have united and no more sperm can enter the egg.

In twelve minutes (in an un-air-conditioned lab in Florida's summer heat), the fertilized egg of *Serranus* starts to divide, and two cells form, then four, eight, and so on, in geometric progression, until in six hours the head of the embryo starts to form. And in eighteen hours a newborn fish, from only one parent, breaks out of its egg membrane and swims free as a larval hermaphrodite.

This feat of sexual independence is an advantage in the survival of a species if an individual gets isolated during the mating season. When I published my finding, Dr. Giles Mead, then with the National Museum in Washington, D.C., wrote and told me he had just found that in a species of deep-sea fish, *Parasudis truculentus,* every adult specimen had an ovotestis capable of producing both eggs and sperm at the same time. Whether self-fertilization actually takes place among these fish may never be demonstrated, but it's a logical solution if such a fish can't find a mate in the abyssal darkness of the ocean bottom.

Recently another remarkable case of self fertilization has been demonstrated in a fresh-water killifish, *Rivulus*. Dr. Robert Harrington of the Entomological Research Center at Vero Beach, Florida, showed that *Rivulus* has its own unique type of hermaphroditism. This fish fertilizes its eggs internally and may delay laying them until the embryo starts to form. "Those fish have a helluva dull sex life—vive la différence!" was Dr. Heller's reaction to the *Rivulus* news.

It took me many more dives, hours of observation in the Lab, and years of work running simultaneously with studies of sharks, rays, other fishes, and non-fish marine animals before I learned the even stranger behavior of *Serranus*, a fish I may never really finish studying or learning all about.

After discovering that each *Serranus* could fertilize its own eggs, I witnessed some surprising activity on a subsequent dive at the red reef near Boca Grande Pass.

I had learned that slack high tide, when the sun is bright, is the best time for diving in murky water in or near the passes. The clear offshore Gulf waters have flooded into the bay and suspended particles have had time to settle. Once the tide starts out, it stirs the muddy bottom of the bays, rich with phosphates and other nutrients, and within minutes water with 30-foot visibility is so murky you can't see your flippers. At the red reef, a slight northward current along the beach of Gasparilla Island catches the newly stirred particles pouring out of Boca Grande Pass and causes the rapid change.

Conditions were against me this particular day. The visibility

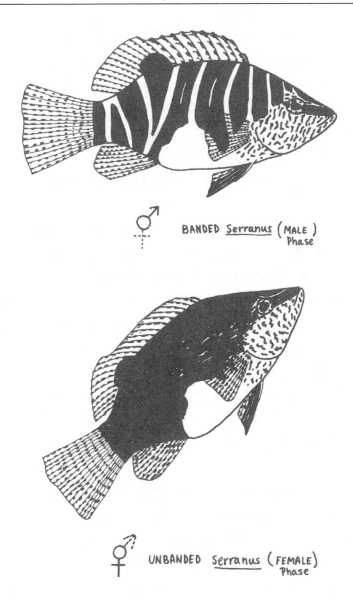

BANDED <u>Serranus</u> (MALE) Phase

UNBANDED <u>Serranus</u> (FEMALE) Phase

was poor, to begin with, and we got a late start. I wasn't ready for the dive until the slack tide was finishing.

The water was murky green at the surface; at the bottom it was brown. But even in water with only two feet of visibility I could find *Serranus*.

An egg-swollen *Serranus* that I tried to steer into my jar started paying less attention to me than to another swollen *Serranus*, and was twisting its body into an S-curve. This is a familiar courtship behavior many fish show. I had seen this when studying the sexual behavior of fresh-water platyfish and swordtails for my doctorate thesis. Fish S-curve also when fighting, especially the more dominant individual. The body twists so that the head goes to one side and the tail to the other.

The second *Serranus* responded to the first with a strong S-curve in the opposite direction. They looked like a pair of male fish working out a hierarchy—beginning a "fight" over territory, nesting, or a female. The "peck order" (which fish is dominant over the other) can be worked out without any actual pecking or biting, but with aggressive intimidating peck-like movements, lunges, or S-curves.

As the two *Serranus* S-curved at each other, the white belly patch on each was very conspicuous and stood out, exaggerating the swollen abdomen on first one side, then the other, as the S was reversed. The bulging bellies were kept toward the partner as the pair chased and rotated around each other. They swam in jerking movements, then suddenly rushed upward, curled around each other, and for an instant seemed glued together; then they snapped apart. I knew I had witnessed their mating. They repeated this pattern several times; then I caught them in my glass jar and took them, along with 40 others I caught before my air tank emptied, back to the Lab.

We kept a large plastic garbage can in the *Dancer* and filled it halfway with sea water and put in a small aerator to keep all the fishes alive until we got them into the Lab's aquariums. The aerator made so many bubbles that we usually just checked to see whether the fish were swimming around O.K. or starting to look sick. But on this day the aerator wasn't working. We put some fish in the garbage can and others into pails of sea water so we wouldn't overcrowd them. As the *Dancer* went back to the Lab, I was watching the *Serranus* in the garbage can, and saw several pairs make that sudden upward swim, adhere, then snap apart.

When we reached the Lab, I sampled the surface of the water in the garbage can and looked at it under the microscope. It was dense with fertilized eggs in early development, with four, eight, and sixteen cell stages.

When we put a group of the newly caught *Serranus* into an aquarium, some started S-curving at others, formed partnerships, and the pairs went through a series of mating snaps. In about an hour this activity stopped, their bellies were flat, and the surface of the aquarium was full of floating eggs that showed up like a string of tiny glass beads across the waterline at the glass front of the tank. The number of stages in egg development corresponded with the number of snaps that had taken place.

The following day, at about the same time, a few matings took place. The third day, the fish no longer swelled with eggs. Like all groupers kept in captivity, they became sterile. Dr. Breder was writing a book on the reproductive behavior of fishes and told me that practically nothing was known about the mating behavior of fishes in the family *Serranidae,* even though groupers, from small species to the giant sea bass, are among the most common animals displayed at public aquariums and oceanariums.

We tried many methods to keep the *Serranus* at the Lab feeling as if they were back in the sea. We built a big plywood box, lined it with fiberglass, filled it with running sea water, put in red rocks and algae from the reef, and tried to reconstruct a natural home for them. Nothing worked. In a few days they flattened out, never to swell again with a fresh batch of eggs unless we put them back in the open sea.

And so we did this. After we kept some *Serranus* at the Lab, we marked them by cutting out a notch in one of their fins and then returned them to the ledge from which we'd caught them. Several days later, when I went back to the ledge, I could find most of our notched fish in the same general area where they had taken up home again. They were ovulating again and mating every day.

Eventually we worked out the mating season, in our area, of this species of *Serranus*. From April, when the water is gradually getting warm, to around mid-September, when cool spells start, these little groupers mate every day. They start ovulating early in the day, and by mating time their bellies are turgid with full-blown eggs. It took awhile to find out what determined the time of day mating would begin. I'm still not sure of all the factors involved, but in the years that followed we learned a great deal from our frequent dives and attempts to photograph them in their natural habitat.

Hera learned to scuba dive with a miniature tank when she was five, and she was with me the first time we were lucky enough to get underwater movies of the mating behavior of these hermaphroditic fish. Fred Logan, a businessman in Sarasota whose hobby is underwater photography, offered to help me. He welcomed an excuse to take a day off from his office. He envied my being able to use scuba diving as a frequent part of my work.

We found an ideal place and day for taking underwater movies. The water was calm and clear, and the bright sun made the red and yellow sponge-encrusted rocks a brilliant background for a color movie. Two days before, I had found this beautiful small reef closer to shore and farther north than the other reef where I'd been working. The water was only about 10 feet deep and the hermaphrodites were numerous. I had observed them mating, so absorbed in their activity once they got started that a diver could ease up to within a foot without disturbing them.

Conditions were even better for photography when Fred Logan, Hera, and I put on our scuba gear and went down to photograph the event. It was the exact time I'd witnessed the mating orgy two days before. When we got to the bottom, the fish were all around us very busy eating and not the least bit interested in sex. Fred and Hera had never seen this mating in nature but I had briefed them on what to expect. "Don't worry," I assured them. "It's so dramatic and distinct—every *Serranus* on the whole reef will be engaged in this S-curving, belly flashing, upward swim, and climactic snap—that you can't miss it."

The three of us settled on the bottom and stretched out quietly on the reef. Fred's camera was all set up and his finger ready over the starter button with the lens pointed in the direction of several *Serranus*, who interrupted their pickings through the debris on the rocks to glance at us occasionally without a trace of fear.

Fred and Hera looked at me through their face masks from time to time, asking with their eyes "Where's the action?" I wanted to use that gesture I learned in Egypt, so handy when underwater with an Arab, of putting the ends of one's five fingers together and drawing downward—the silent equivalent of "Stanishwaya" or "Wait a bit." Hera so persistently nagged at my shoulder strap (at the age she was then, I always kept her next to me when she scuba dived) that I finally put her back in the boat.

Fred and I remained watching. Fred was patient a long time. The little groupers were plump but not fully swollen. Fred took some general pictures of the scenery and of the fishes feeding and then motioned he wanted to go up and change film. Even in the warm water, we were cold by that time. The air in our tanks was running low, and the warm sun directly on our backs felt good when we got into the boat. We ate lunch and sunbathed.

Every ten minutes or so, I free-dived down and looked at the fish. They were still eating, and although I never saw one swallow a succulent blenny or a pistol shrimp, which they particularly like, their bellies were getting noticeably fuller. If they were getting much food, picking at the sponges and algae, the particles must have been practically microscopic. I couldn't see what they were eating even with my face very close to them and with the magnifying effect of my face mask. Their bellies had become larger in the two hours we'd been watching them.

Then suddenly the action started. In the next fifteen minutes, before we ran out of air, we got the film we wanted. The water had begun to grow murky and less calm as the ebbing tide from Gasparilla Sound brought out the muddy bay water, but Fred was able to get close enough to the fish to get clear pictures.

I learned that the tidal cycle influences the mating time of these fish, at least the ones we studied that lived close to shore where the tidal effect on bottom-living sea animals is more strongly felt. I was finally able to calculate to within a few minutes when mating would start in any one of the several places where we dived to study them. Mating time varies with the depth of water, rate of flow, nearness to an inlet pass, size of inlet, etc., but in general it is within a half to one and a half hours after high tide—just after the still slack tide and clear water, so perfect for skin diving, starts running out to sea and stirs up the mud of the shallow bays. This same action in the environment seems to stir *Serranus*, and its increased picking movements, without much if any actual eating, may reflect the disturbance within its body as the eggs, ready to be fertilized, break out of their follicles and swell. The swollen conspicuous white bellies are probably a factor in triggering off the start of courtship in which belly flashing plays an important role in stimulating actual mating.

Our studies in the Laboratory, watching fish we had caught earlier in the day, revealed extraordinary details of the mating behavior of these hermaphrodites. Each adult individual can act as a male or a female or both simultaneously, depending on the situation.

We found that during one mating sequence ending in the snap (the moment when eggs and sperm are released), one fish leads and the other follows. The fish that follows is, as might be expected, playing the male role. It then has bold, broad, dark vertical hands on its body. The fish in the female phase is unbanded, often has the greater abdominal enlargement, and takes the lead during the final upward swim. These two color phases are helpful to the observer, because the fish are capable of reversing their roles within a matter of seconds. When not spawning, their color patterns vary from the banded to the unbanded stage. The former is the more common, especially in young fish.

In the Lab, we would put two hermaphrodites together when both were filled with eggs and unbanded. After what appeared to be a frustrating attempt at courtship by two females, the fish began to lunge and make pecking movements at each other; S-curving and display of the white abdomen intensified between pecks until one fish, usually the larger, managed to force the other into a corner

My first meeting with Jacques Cousteau in Miami, where we delivered a joint lecture on sharks, led to my guest appearance on the first of his TV series, "The Underwater World of Jacques Cousteau".

until it appeared to give up. The smaller one then darkened into the banded male phase, came out of its corner, and pursued the larger fish in the typical courtship maneuvers. Soon the abdomen of the larger fish grew noticeably flatter, as it expelled the eggs, while the other fish was still swollen. Then sex roles reversed. Several reversals may occur before the spawning period ends.

The introduction of a third fish into an aquarium with a spawning pair can also cause a sudden sex reversal. In one situation, spawning started between a pair after the smaller had been forced into the male role. The third fish, the largest, had just spawned as a female in another aquarium, and had released all or most of its eggs. When introduced into the

My first diving apparatus, the Cousteau designed aqua lung. Serranus captured in the jar.

aquarium, it was attracted to the smaller of the spawning pair, which had the more swollen abdomen. This small male-phase fish rapidly lost its bands, switched to the role of a female, ignored its previous mate, with whom it had snapped as a male a few seconds earlier, and snapped as a female with the third fish. The time between the two snaps was less than a minute.

In a crowded aquarium, several fish may take part in a snap. The leader in the chase, in the unbanded female phase, seems to control the climax of the upward swim. Five or six "males" may dart upward with "her"—all but one making a lone snap without touching "her."

We made a surprising discovery when studying photographs taken just before the snap. At this critical moment, the leading female-phase fish became banded! At first this confused the otherwise clear-cut courtship relationship between an unbanded female phase and a banded male phase, until I realized that the banding on the leading fish is the reverse, or negative, of that on the male-phase fish.

We still have a lot to learn about the unusual reproductive habits of this hermaphrodite. How can it reverse its sexual roles so quickly? What controls the color changes? We are still not sure if the fish playing the male role releases a few eggs at the time the sperm are ejaculated, or if the fish in the female role is capable of holding back its sperm supply. There are definite advantages in crossbreeding, and it appears that self-fertilization is only used as an emergency method when a mate is not available. However, a good technique has yet to be developed to discover whether or not self-fertilized eggs ever result from the spawning of a pair of these odd, hermaphroditic fish.

5   Blennies and Mazes

THE MAIN PROJECTS we started during the first two years of the Cape Haze Marine Lab's operation were establishing a preserved reference collection of the local marine life and our studies of sharks and hermaphroditic groupers. While diving for *Serranus*, I had collected five different kinds of blennies in glass jars, but I had seen other species I had yet to catch. During a lull in our activities one fine day in 1956, Beryl and I decided to make a special effort to complete the Lab's reference collection of local blennies.

The saw was Beryl's idea, yet as he steered the *Dancer* toward Stump Pass he suddenly gave his cap a push forward and shook his head. "You know," he said, looking at me as if he had suddenly

Nichols' Blenny

discovered himself, "after twenty-seven years of all kinds of fishing, I'll be damned if this isn't the first time I'm going fishing with a saw!"

The blennies we were after live in sunken tree stumps in Stump Pass, where Lemon Bay connects with the Gulf of Mexico. We came across this wonderful collecting spot for blennies quite by accident. Beryl's son Larry and I were skin diving in Stump Pass one warm day at slack high tide, that ideal time to dive. We watched schools of mullet feeding on plankton near the surface and large snook resting on the sandy bottom. Sheepshead, spadefish, and triggerfish slipped in and out among the roots of the big stumps, playing hide-and-seek with us. Two porpoises swam by in the deeper part of the channel and gave us the feeling that we were swimming in the big fish tank at Marineland. Larry pointed his finger at a thick horizontal branch of a stump. I noticed hundreds of tiny eyes watching us. Some of the eyes had tentacles sticking out over them, making them look like grasshopper's eyes. Only the heads of the fish showed, propped up on pelvic fins that serve as forearms. The rest of the body was backed down a hole in the stump. The whole branch of the stump was perforated with irregular rows of holes, made originally by boring mollusks, and almost every hole had been taken up by a blenny. It reminded me of a block of five-story brownstone buildings on the East Side of New York on a hot summer afternoon, when all the windows are ajar and almost every window has a face, propped on elbows, looking out into the street below. The population of blennies at Stump Pass was as dense as on the red reef off Boca Grande.

That first day we caught only a few. Then Beryl came up with the idea of sawing off the whole branch and shaking the blennies out into a tub of sea water in our boat.

As the Cape Haze Marine Laboratory was a place where visiting scientists could come to work on research problems dealing with marine biology, it was important that we be able to tell them the exact identification of all the local marine life. There was no one book on the fishes of the west coast of Florida. Scattered scientific papers gave bits of information. It was comparatively easy to identify the 30 or so common food and game fishes. But what about the 200 or more different kinds of fishes our collecting efforts were turning up? These were presenting problems—blennies, for example.

Blennies are small fish, usually under three inches. They are poor swimmers and are better at crawling and jumping. In an aquarium, they make nice pets and live contentedly in shells or artificial contraptions with openings they can back into and just leave their heads out to keep track of what's going on outside. Occasionally they come out of their holes to visit, fight, or flirt with a neighbor or to forage for food. They become accustomed to a hand that feeds them and will swim clumsily to the top of the aquarium to take a bit of food from your fingers. They seem to know their surrounding terrain and scutter quickly back to their home hole when frightened, usually with their bodies close to the substrate, rather than swimming through the water, even if it takes longer. Sometimes they follow a complex labyrinth of alleyways to get back to their favorite hiding hole, jumping fences along the way like a petty thief in his own territory evading a chasing policeman.

While skin diving at Stump Pass, I got to recognize an individual blenny with a scar on his head who lived in a hole with a sponge growth just above. I captured him by poking a piece of wire down the hole and catching him in my glass jar as he came out. Then I carried him several feet away from his hole and released him. In no time he was back in his hole. I repeated this several times, taking him farther and farther away from his tiny den until I was releasing him nearly twenty feet away. I followed him on his trip home, floating at the surface and breathing through my snorkle. I could see him nervously hurrying along the logs and stones and across stretches of sand. Every once in a while, he seemed to be frightened by my paddling movements five or six feet above him, and he would pause and back down another hole, usually to come out again followed by the head of another blenny who looked like a bouncer.

It occurred to me that blennies might make excellent subjects to test in a maze. Few studies had been made on fish to test their ability to learn the patterns of a maze, but I felt the blennies at Stump Pass could learn to run through a complex maze with ease. I hoped someday to work on this problem, but I have not yet been able to do so.

Dr. Lester Aronson, the Curator of Animal Behavior at the American Museum of Natural History, was working on a problem at the Lab dealing with orientation and jumping behavior of gobies. He also became interested in the blennies at Stump Pass. "Let's test a whole series of blennies and get some real data on their homing ability," he suggested. One of the technical difficulties involved was how to keep track of the individual blennies and their home holes. We finally marked the holes by placing pins with different-colored heads a short distance above each hole after we had captured its inhabitant. Then we stitched a small piece of my silk embroidery thread, of a color to match the pinhead, through a flap of skin on the head of the blenny. Then Lester would wait near the home hole while I would swim off a measured distance with the marked blenny and then release it, and at the same time stick my head out of the water and call out "Now!" to Beryl, who was in the boat with a stopwatch. We were even toying with the idea of running some of these tests after dark with an underwater flashlight to see if there was any difference in behavior at night. But we never attempted this. It was bad enough during the day, but we would have attracted more attention at night from the fishermen passing by in boats. It would have taken just too much explaining to be caught with Beryl holding a gasoline lamp over our heads while Dr. Aronson and I were sewing colored embroidery threads onto the heads of blennies. The results of our first tests were not as clearcut as we had hoped. No blenny was as deft and accurate in finding his home hole as the first scared blenny I had tested. Perhaps we handled the blennies too much when sewing the threads on them. Perhaps they didn't like to go back to a hole with a colored marker above it. Perhaps the first blenny I tested was an exception. We're a long way from finding out about blenny homing behavior, but it is one of the many intriguing problems we keep in mind.

For our museum collections, we wanted preserved records of every kind of fish in our area. Small specimens fit best in our jars. We made a series of various young stages to study the changes in development. Beryl, after years of always being proud to catch a particularly large fish, would come in beaming and boast, "I'll bet I've got the smallest snook anyone around here can catch." With a one-man fine-mesh push net and other collecting gear, Beryl would try out different parts of the bay where the sand and mud bottom changed to shell or to some special plant growth, or a place where sponges were plentiful or where anemones burrowed into deep tunnels in the mud. He would bring back tiny pipefishes and gobies less than an inch long or a beautiful young angelfish. It delighted him to find many kinds of little animals new to him hiding around the waters of the Gasparilla area, where he had fished for so many years and which he thought he knew inside out. He was a keen observer and had the curiosity and enthusiasm about marine life that makes a good naturalist. He became fascinated with plankton after we made our first plankton haul, and this new world opened up to him through the microscope.

Beryl was good at many things and had wide and practical knowledge and experience. It was inevitable that the higher salaries of other jobs would lure him away from the Laboratory, especially when his three children had reached high school age. He understandably took an offer as head of the water company in Placida and supervised the building of a dam. But I believe his heart will always be in the fishing business.

Before leaving blennies, I must tell you about the close call Nichols' blenny or *Blennius nicholsi* had to being named *Hypsoblennnis gallowayi*. It will give you some idea what is behind the simple ichthyological statement (a taxonomic conclusion I was forced to make) that *Hypsoblennius gallowayi* Fowler = *Blennius nicholsi* Tavolga.

In a scientific name, the first word refers to the genus (often changed), the second to the species (which only under exceptional circumstance can be changed if it is the first description), and the third to the authority's name—the person who described this species.

When I first caught the rare Nichols' blenny in my glass jar, I couldn't identify it. While I was trying to figure out what species it was, a reprint of an article by Dr. William Tavolga arrived in the mail. Bill Tavolga, now an authority on fish behavior and known all over the world for his studies of fish sounds, I remember as a skinny young graduate student at New York University with tremendous energy and enthusiasm for his studies. The biology courses we shared seemed to burn more calories than he ever took time to replace. All his friends were relieved when he married another biology major. Whatever the Danish expression for *gemütlich* is, it applies to her. Margaret, who was as excellent a cook as she was a biology student, saved Bill from vanishing away to nothing. She deserves much of the credit for the fine contributions her husband has since made in the field of zoology, while becoming an expert herself on the behavior of dolphins.

So Bill has written a paper on fish taxonomy, I thought as I saw the title of his paper and started reading about a new species of fish he had found near Marineland on the east coast of Florida. I was pleased to see that Bill had named the fish after John T. Nichols, the lovable ichthyologist at the American Museum of Natural History. From a drawing on the next page I immediately recognized the odd blenny I had just caught in the sunken logs at Stump Pass. I was amazed by the coincidence and dropped a note to Bill telling him that the new species he had just described, known only from three preserved specimens, was swimming around in the Lab aquarium. This was the start of some correspondence that led Bill and Margaret to come to the Lab to study sounds of blennies and other fishes.

A few weeks after I received Bill's reprint, I heard about *Hypsoblennius gallowayi.*

Mr. Galloway wrote to me at the Lab, congratulating us on our work. He and his wife were amateur naturalists, he wrote, and they collected plants and animals, alive, pickled, pressed, or fossilized. Could he send me marine specimens from time to time that he couldn't identify? He and his wife wanted very much to visit the Lab, but because they were getting too advanced in years it was difficult for them to make the trip from Punta Gorda (20 miles away). They hoped I would visit them.

Whenever I identified specimens for Mr. Galloway, he credited me in his weekly news column for the *Punta Gorda Herald*; and from his comments on marine life, I realized he was a person with unusual insight into nature and someone I would like to meet. When I went to see the Galloways, I was unprepared for the tenderness of this old couple and the contents of their house. Mrs. Galloway was blind but she cooked and took care of her husband, who moved with effort but insisted on taking me through every room. The necessities of the home (eating, sleeping, and washing facilities) were squeezed into corners of the house, which was primarily a museum of a tremendous variety of specimens. Mr. Galloway was also a taxidermist and an artist. He showed me folios of his paintings of plants and plant parts in fine anatomical detail. He had boxes of archaeological specimens he had dug up and collected locally, thousands of stuffed and preserved mammals, birds, reptiles, amphibians, and fishes he had collected himself from all over the world. Unusual invertebrates cluttered corners of his shelves. He had a huge collection of fishes in bottles of formalin, but I was dismayed that most of the specimens lacked labels. The identification of a valuable specimen can be made at any time, but the place and date of collection should be placed on a specimen as soon as possible. These data Mr. Galloway carried in his head, though he said he was hoping to get around to labeling one of these days.

His prize fish specimens, he said, were not in his house but at the Philadelphia Academy of Sciences. He had discovered a new species of blenny. He said it was collected locally by a young man who brought two dead specimens to Mr. Galloway, not knowing what kind of fish they were. Mr. Galloway knew they were a kind of blenny, but couldn't identify the exact species, so he sent it up to Henry W. Fowler, who declared it a new species.

Mr. Fowler was probably the most voluminous ichthyological writer the world has known. I met him once for a short time when I visited the Philadelphia Academy of Sciences years before. A number of his volumes sat on my shelves with pages worn from the many times I referred to them. Mr. Fowler's works cover both fresh water and marine fishes from every part of the world. He must have described close to a thousand "new species," most of which he later pointed out himself were not new but the same as previous species he had described. His publications are valuable for reviewing all the difficult literature in foreign languages and ancient journals bearing on the description of a species and its range. But the quality of his descriptions suffers from the quantity and pace at which he wrote. He did not take time to check his descriptions carefully.

I was curious to see what this Galloway blenny was like, so I wrote to Mr. Fowler for a copy of his publication describing it, which he kindly sent by return mail. His illustration of the new species looked unlike any blenny I knew. His written description and measurements seemed very much like Dr. Tavolga's *Blennius nicholsi,* but did not mention the two enlarged canine teeth, especially in males, characteristic of the species of blennies grouped in the genus *Blennius.* Mr. Fowler had described the Galloway blenny in a new genus, *Hypsoblennius* (referring to the high dorsal fin). He named the new species in honor of the sweet old couple.

The Galloways were nature lovers who studied quietly and informally for sheer enjoyment. We carried on a correspondence for two years until a cheery Christmas card came, signed from both of them with a note from Mr. Galloway saying that his wife had passed away, but before she did she told him to send her greetings in any case. There was no tone of sadness in the card or note, and I marveled at the placid and cheerful attitude toward phases of life and death this couple had achieved, and wondered if I would ever get as close to understanding and accepting nature as they did. Mr. Galloway's death was announced in the papers shortly after that.

I then learned that the week after I visited him Mr. Galloway drew up a will leaving all his collection and paintings to the Laboratory. It was more than the small Lab rooms could contain, and so much of it did not pertain to marine biology. I was relieved

The blenny maze designed by Beryl Chadwick.

and happy to hear that the citizens of Punta Gorda wished to set up a Galloway Memorial Museum, and the public now can see this unique collection in Punta Gorda.

Still the blenny story hadn't ended. One day a young man came to the Lab, excitedly introduced himself as James Stephens, and reported that he had brought us alive a new species of blenny. "How do you know they're a new species?" I asked. "I caught some once long ago and gave them to Mr. Galloway, who told me later that a scientist up north said they were definitely a new species. This is the first time I could catch them alive."

I was also excited to see this fish, as I had been watching for this rare blenny Mr. Fowler named after Mr. Galloway, but in several years of collecting in the area where it was found I had never seen one.

We poured the contents of the tin can, which Mr. Stephens brought in from his car, into an aquarium, and I got a good look at the fish. There was no doubt about it, these fish looked exactly the same as the blenny we called *Blennius nicholsi* according to Dr. Tavolga. I netted one, wrapped it in wet cotton, and quickly looked at its teeth under a microscope. The two enlarged canine teeth were there.

I wrote to a young skin-diving ichthyologist, an excellent and careful taxonomist, Dr. James Bölke, who also worked at the Philadelphia Academy of Sciences, and asked him if he would examine the type specimen on which Mr. Fowler had based his description of the new genus and species, *Hypsoblennius gallowayi*, and see if Mr. Fowler had overlooked the two canine teeth. He had, and because Dr. Tavolga's paper appeared a few months earlier than the publication of Mr. Fowler's paper, the name *Blennius nicholsi* has priority over the name *Hypsoblennius gallowayi*.

6 Sharks That Ring Bells

AS THE LAB BECAME BETTER KNOWN, the problem of unwelcome visitors arose. Once we found an eleven-foot tiger shark dead in our pen with the hoop of a large, long-handled dip net jammed over its gills. This was a particularly handsome shark that was living well in our pen and eating huge quantities of shark meat left over from our routine dissections. We couldn't imagine how the naked wire hoop of the dip net, which Beryl had left on the dock until he had time to put a new net bag on it, got into the pen and over the shark's head.

Beryl learned how when he was getting a haircut. The man in the next chair was telling his barber.

"That Cape Haze Marine Lab is a dangerous place," he said, and went on to tell how he and his wife went down to look around one day, but nobody was there and the buildings were locked up. They were fascinated by the large tiger shark in the pen and, finding the big hoop with the long handle, decided to see if they could make the shark swim through the hoop. As the shark swam close to the dock, the man's wife held on to the end of the eight-foot handle and lowered the hoop directly in the shark's path. The shark swam into it, but the pectoral fins prevented the hoop from getting further than the gill region. "That goddam shark was ready to pull my wife into the water if she hadn't let go the handle," he told the barber. They then hurried away from the "dangerous" Lab, leaving the tiger shark to thrash itself to death.

Beryl and I decided we had to take more precautions, and as much as we disliked putting up eyesores, we made "No Trespassing" signs and put a chain across the driveway to the Lab. This discouraged most off-hour visitors, but not the worst kind. As the Lab grew, we had to put up stronger warning signs and barbed-wire fencing. Even then, I sometimes caught trespassers in the act when I was working late at the Lab or on a Sunday. One Sunday, I looked up from my desk and was horrified to see a child about four years old sitting on the feeding platform dangling his feet in the shark pen. His parents had brought him. They had ignored the signs and climbed the fence. They saw no sharks in the pen and were wandering about the grounds. An eleven-foot dusky shark was swimming in the deeper murky water of the pen. Fortunately, it had eaten its fill the day before.

One evening, when I had to work very late, I watched as a man, pushing down one barbed wire with his foot and lifting the upper wire with his hand, allowed a pretty girl to step through. I caught up with them as they neared the shark pen and he was saying, "Wait till you see this!" I interrupted their adventure and talked with them. He defiantly told me, "We weren't going to harm your precious sharks." I might have lost my temper except I was stalling for time. I had already called the police.

The couple was arrested. The arrest, announced in the paper the next day, was enough to keep trespassers away for some time.

Another time, a man with three young ladies in his boat pulled up alongside a small pen which held twenty baby sharks we had delivered the day before and were hoping would live. The man began showing off by catching one of the two-foot sharks by the tail, twirling it around his head, and hurling it across the pen while his audience giggled. He quickly started up his motor when he saw me coming out of the Lab.

We continued setting our lines for sharks. After a day of diving on the reefs, Beryl would check the shark line on the way back and we might return to the Lab with garbage cans of two-inch gobies, blennies, and serranids, and a twelve-foot shark. I kept on experimenting and studying abdominal pores and anything else I could learn about the sharks we caught. This led to a series of experiments that ultimately got too much publicity and stuck me with the dubious label of "shark lady."

Somehow my reputation of diving and using a spear to collect fishes (described in my book, *Lady with a Spear*) made reporters and writers of magazine articles assume that I collected big sharks by diving and spearing them. Even when I corrected this notion in an interview and stipulated before the interview that I be allowed to see the text of the article before it was sent to press (something popular magazine writers and editors hate to do), many bloopers got through. In one case, the writer carefully avoided any hint that I dived with and speared the sharks I worked with, but when the art department of the magazine got to work they sketched a picture of me diving, with a one-rubber arbalete spear gun that would barely hold a mullet, about to shoot the spear into a gigantic shark. This came out as a two-page spread in the magazine.

The things we were learning about sharks and our success in keeping sharks in captivity drew the attention of scientists from various parts of the world who were interested in problems related to sharks. They came from Africa, Germany, Italy, Israel, England, France, Denmark, and Japan to work with us. We were able to expand our research with additional financial help we got in the form of grants from the National Science Foundation, the Office of Naval Research, the American Philosophical Society, and the Selby Foundation.

There are about 250* different species of sharks living in the world, ranging from a cute little deep-water species less than a foot long to the giant whale shark reported to attain 60 feet in length. The largest shark that ever lived was *Carcharodon megalodon.* Its black-and-gray fossilized teeth, measuring up to six inches, are commonly found on Florida beaches. It probably grew to nearly 100 feet in length and is related to the most dangerous man-eating shark now living, the great white shark, *Carcharodon carcharias.* Fortunately, the white shark grows to only 20 feet.

The wide shallow continental shelf extending about 100 miles out on the west coast of Florida limits the variety of sharks we can catch to about 18 species. For practical reasons, we had to operate

---

* Editor's note: Now, in 2009, almost 500 different species of sharks have been identified (Dr. Jose Castro, personal communication, October, 2009).

our shark lines within eight miles of shore and had little chance of catching real deep-water or oceanic sharks. We caught three different species of hammerhead sharks, two species of blackfin sharks, several types of small dogfish sharks, and on rare occasions a white shark or a sand-tiger shark. But we could count on a good supply of dusky and sandbar sharks each winter, nurse sharks and tiger sharks all year round, and many bull and lemon sharks during the spring, summer, and fall.

Most of the sharks we caught were from 5 to 11 feet long. We measured a few dusky, tiger, and great hammerhead sharks that got into the 12- and 14-foot range and nearly 1,000 pounds, but this size group is uncommon in our records.

Our examinations of the contents of hundreds of sharks' stomachs showed us that our local sharks eat over 40 varieties of fish, including eels, some prickly bony fishes (e.g., spiny blowfish, and poisonous spined catfish), sting rays, and other sharks. We also often found remains of squid, octopus, shellfish, starfish, crabs, and shrimps, occasionally sea turtles and sea birds, and rarely some porpoise remains. Once we found a yellow-bellied cuckoo and another time a gull with a metal band (which was reported to the amazed bander) in the stomach of the omnivorous tiger shark.

Most of the species we studied seem to mate in the spring and early summer. At this time, the claspers of mature males were red and swollen. Often sperm oozed out of the urogenital papilla,* and on dissection we could see the testes were much enlarged. When we opened the female, we could find wriggling sperm by examining under the microscope smears taken from the vagina, uterus, and

---

* During copulation, sperm from the urogenital papilla enters a groove in the clasper, which swings into a forward position and inserted into the female.

tubes leading toward the ovary. On the ovary, large yellow ripe round eggs were nearly ready to break out of their follicles. Shortly after the mating season, the ovary would contain only small white eggs but the uterus would contain big yellow oval-shaped eggs. These had been fertilized by sperm at some point high in the tubes, and partway down the tubes in the nidamental or shell gland had acquired a soft golden "shell" before settling into spongy pockets in the uterus. Sometimes we found each oval egg had a tiny wriggling embryo on the surface of the yolk.

By plotting the size of these embryos at different months of the year, we could draw their growth curve and determine the gestation period, which varied from 8 to 14 months, depending on the species. The females of most species we studied seem to have a litter of pups only once in two or three years.

The size of the litter depends to some extent on the species of shark. The sand-tiger shark has only two young at a time, and these, according to Stew Springer, start feeding *before they are born* on eggs the mother continues to supply from the ovary. The bull, dusky, sandbar, and lemon sharks have from 5 to 17 young at a time; the great hammerhead can have several dozen, and the tiger shark over a hundred. The older the mother shark becomes, the greater the number of young she carries when pregnant. We also observed a few "senile" big sharks whose ovaries had apparently stopped functioning.

Our shark dissections, especially on pregnant females, supplied fascinating information on the reproductive habits of some large species of sharks about which little was known previously. The most interesting part of our work with sharks was studying the live animal, working with one individual day after day for long periods of time, and getting to know its personality.

One of the most extraordinary sharks I ever got to know was one I gave up for dead the day we caught her in May, 1958. She was an adult lemon shark nearly nine feet long. This was before we devised better methods for bringing live sharks from our setlines in the Gulf back to the Lab. We towed this shark by the hook she had taken in her mouth and firmly set in her jaw. It was a rough day. She was bounced on large waves in the Gulf, and once we got her in the bay, we had to drag her across shallow barnacle flats because of the low tide. All the time, her mouth was being held open by the hook and towline, and water poured into her.

She struggled weakly when we first found her caught on the line. By the time we had her in the large fish pen alongside the Lab's dock, she appeared dead. We tried to revive her by "walking" her around the shallow end of the pen— pushing and pulling her gently so that the water would flow over her gills—but she didn't respond.

We finally gave up. It was too late in the day to hoist her onto the dock to weigh, measure, and dissect her. So we tied her limp form alongside the dock, still with the hook and side line in her mouth, and left, intending to work on her in the morning.

When we saw her next, she was a very lively shark, raising a commotion at the end of the dock. We put her into the pen again, in shallow water where she couldn't struggle too much. We held her down, propped her mouth open with a block of wood, and removed the hook. We had to cut the hook wound open a little wider, in order to back out the barbed hook with long-handled pliers. As soon as we released her, she swam vigorously around the pen and then slowed down to an effortless glide. Only gentle undulating movements of the end of her body and tail were needed to propel her powerful, streamlined body.

So we began to learn how hardy lemon sharks are, unlike other active sharks, and how well they can adjust to captivity. In the years that followed, we found it easy to bring lemon sharks in from our lines in good condition.

After the shark had spent a day in captivity, we discovered the probable reason for her weakness on the line—seven newborn lemon sharks, each about two feet long, swimming with her in the pen.

It was the first time a shark had given birth naturally at the Lab. We had removed many living embryos and near full term pups from gravid dusky and sandbar female sharks we dissected. Those babies, even when we tried "incubator" methods, seldom lived more than a few days. But we had great hopes for raising the newborn lemon sharks.

The day after they were born, they started feeding. The female and her pups had our large 40- by 70-foot pen all to themselves. In a few weeks, they were eating well and the babies grew noticeably fatter. The sight of a person walking to the corner of the shallow end of the pen, where the young sharks were fed, attracted them almost immediately, and the first piece of cut fish dropped into the water set up a feeding frenzy as the seven young sharks crisscrossed each other and rushed around the feeding area.

Then I made a mistake. Our setlines caught a large male lemon shark, and I thought he would make a good mate for the female. We put him in the pen and he did get along fine with her. But the baby sharks began disappearing one by one. Those healthy young sharks, which would have been ideal for testing the function of abdominal pores because of their small size, were soon all gone.

We were left, however, with a handsome pair of adult lemon sharks that were very responsive to signs of food. We had an excellent opportunity to study their feeding behavior.

At first we threw fish (whole mullet) to them. We used so much mullet to bait our shark lines and to feed the sharks we kept alive that we invested in a storage freezer. It paid for itself with the money we saved buying mullet wholesale, 200 pounds at a time.

In order to watch more closely when the sharks were feeding, we built a feeding platform close to the water. We weighed the amount of food we gave them each day, since I found out that practically nothing was known about the amount of food needed to keep a large shark alive. We could make the sharks come to the edge of the platform and take mullet handed to them. Occasionally a shark would miss and a mullet dropped to the bottom, especially when the shark charged toward the platform with open mouth and the feeder understandably let go of the mullet a fraction of a second early. Then we couldn't be sure which shark got the food.

A bright junior high school student, Tommy Romans, was a regular visitor to our lab and begged us to let him be a volunteer worker and, among other things, to let him feed the sharks. The risk we ran having an "attractive hazard" was unavoidable for our studies, but we could avoid letting a fourteen-year-old boy feed sharks from a platform suspended over the pen. Tommy was very persuasive. It could have been annoying except that he was so pleasant. He started working for us, helping in the weighing, measuring, and dissection of dead sharks. He also looked around for any odd jobs he could do, was quick to grab the hose and scrub down the dock when a dissection was over, sweep out the Lab floors, wash our aquariums, empty garbage, help bait hooks on the boat—always showing almost the same enthusiasm and willingness to do these jobs as to join us skin diving. He was a fine diver and loved helping collect small fishes and pulling up the shark line, especially when something was tugging on it. He fed all the little gobies, blennies, *Serranus*, and other animals in the aquarium room, and soon took over weighing and preparing the food for the big sharks.

He watched us feed the sharks. One shark, making a clumsy pass at a mullet, bumped into the feeding platform, which at high tide was at water level. (When the tide was extra high, we didn't feed the sharks. A foot could be mistaken for a mullet.) The mullet fell down, and both sharks went after it. The sharks never fought over the food, but even their gentlest movements around the feeding platform stirred up the soft muck on the bottom of the pen, and you could see down only 2 or 3 feet of the 7-foot depth. We couldn't tell which of the two sharks had got the mullet. Also, we now had a big nurse shark in the pen which stayed on the bottom most of the time.

"Why don't we tie a weak string on the mullet until the shark takes it?" suggested Tommy. It was a good idea, and Tommy undertook the task of tying each mullet with a string several feet long and wrapping the length neatly around the fish, so that it could be thrown to the shark like confetti and the strings wouldn't tangle in the bucket of fish. The sharks came quickly to the splash at the surface, and the mullet could then be pulled into a convenient position for camera studies before a shark was allowed to take the food. This method took away the hazard of feeding the sharks by hand; the records of food consumed by each shark became more accurate; and Tommy had won himself the job of feeding them. His interest went deeper. He began borrowing books from our library to take home, and would discuss them with me the next day. I wondered if he was ever getting his regular school homework done.

Dr. Lester Aronson, an expert in animal psychology, came to the Lab during the early summer of 1958, "Has anyone ever made a study of the learning behavior of sharks?" I asked him.

Dr. Aronson told us that some experiments on the olfactory sense of small sharks had been done. Plugging a nostril will cause a dogfish to swim in circles toward the unplugged side if a scent of food is in the water. But no sophisticated experiments of the kind done with birds and rats had been tried on sharks. Sharks, being such primitive fish with a primitive brain and poorly developed visual apparatus, were generally considered rather stupid—poor subjects for the classical experiments done with higher animals. "Besides," he said, "they're difficult to keep as experimental animals, and no one has tried putting them into a Skinner box!"

"Our sharks are *smart*, and, boy, can they *see* us coming with the food!" Tommy said, defending his pets.

Dr. Aronson then encouraged us. "You certainly have a unique and good setup for testing their ability to learn some simple task."

We prompted Dr. Aronson further and before the day was over we had a plan. He suggested that we place a target in the water and train the shark to take mullet from the target in such a way that the shark would bump its nose against it, which would ring a bell. After a long training period, we might condition a shark to associate the target with food, and to press the target and ring the bell even when the food was no longer presented along with the target.

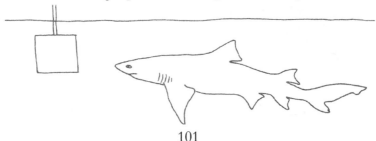

Before he returned to New York, Dr. Aronson helped us work out a design for the apparatus to try an experiment in "instrumental conditioning," using the feeding end of the shark pen as a kind of Skinner box.

"I wish I could stay longer and help try to train those sharks. They do look smart. Don't be discouraged. It may take months, but it really will be something if you can succeed."

At the end of the summer, Tommy hated to leave the Lab. Oley Farver, another ex-commercial fisherman, had started working with us. Oley, with Tommy's assistance, had built a longer, stronger platform with a railing. He also built a special arm to which the target, a 16- by 16-inch plywood square painted white, could be attached during feeding tests. The target could be placed so that it was just below the water surface regardless of the tide, and a firm push on it would close an electrical circuit, causing a submerged bell (an ordinary 6-volt doorbell, sealed in a metal cylinder) to ring.

Near the end of September, we started a strict training program for the sharks. Every day at 3:00 P.M., Oley and I put the target in the water for a maximum of 20 minutes and fed the sharks in front of it, dangling the mullet from a string. At first, the sharks seemed wary of the target and hesitated to take the food. But after a few days they could be lured in so close that in order to get the food into its mouth, each shark was forced to press its nose against the target. We increased the intensity of the bell from a weak buzz the first week to a loud ring that could be heard clearly above the water by the end of the second week.

From our feeding records, we were surprised to learn that it took only 15 pounds of mullet a week to keep a full-grown lemon shark of nine feet healthy and active. If we gave them all they could eat, each would take more than 30 pounds at one time, but then they fed poorly or refused food for days. A shark in its natural habitat probably goes weeks, or even months during cold weather, without eating and then may find a huge meal, such as a weakened porpoise or sea turtle, that will keep it going for another long period without food.

In order to keep our lemon sharks in training every day, and to make them repeat pressing the target and ringing the bell as many times as possible each day, we cut the mullet into smaller and smaller pieces until each shark had to ring the bell about 6 to 10 times to get his daily quota of two pounds.

In a few weeks, the sharks were taking food from the target so rapidly that the daily training test was often over in less than 6 minutes. We had to devise a fast way to reload food on the target, for when one shark removed the food the other shark was often directly behind. We attached a thin wire to the center of the target. To each piece of food we then attached a short loop of weak string. We looped all the pieces of food on the end of the wire that reached the platform. As a shark pressed the target and removed the food, the next piece could be released to slide down the wire and in place on the target as the second shark came in.

We could see clearly how wrong was the old belief that a shark has to roll over to take a bite. Even though the mouth is on the underside of its head, the lemon sharks had no difficulty coming squarely at the target and removing the food without rolling. The snout projecting forward over the mouth, however, did make it necessary for the end of the snout to give the target a good bump in the process.

The large nurse shark we also had in the pen occasionally gummed up the training period. It wasn't satisfied with the food we gave it in another part of the pen. It wanted to get in on the act. But the nurse shark, *Ginglymostoma cirratum*, is quite different from the lemon shark, *Negaprion brevivostris*. It belongs to another family of awkward-looking sharks and is slower, more sluggish, and clumsy; it spends most of its time lying on the bottom, often under ledges or in dark caves. The adults are about the same size as lemon sharks (9 feet), but have tiny eyes with much poorer vision. Our later tests on baby nurse sharks proved they have good vision. But the huge nurse shark in the pen with the swift- feeding lemon sharks didn't seem to be able to see the target. Slowly it would pick up the scent of food and, unlike the lemons, could dangle in front of the target in a vertical position, holding this position by waving its broad fanlike pectoral (arm) fins while gently passing its small-toothed, blubber-lipped mouth over the surface of the target. It would smell and feel its way to the food with the help of its barbels, a pair of whiskers on either side of its mouth, and then, by closing its gill slits and opening its mouth, create a vacuum that would suck in the food and break the string, stealing the lemon shark's food without ringing the bell. The nurse shark got in the way so much that sometimes we'd have to poke him on the nose with a stick to get him away from the target so we could get on with the training of the lemons.

After six weeks of feeding the lemon sharks from the target, we gave them the big test. We put the target in the water at the appointed time but with no food on it. The male lemon shark who usually responded first rushed at the target with his mouth open, then swerved aside when he reached it and found no food. The second time, and for eight more times, he came in slowly looking over the target without touching it, a few times brushing it lightly. Finally, he nuzzled the empty target hard enough to set off the automatic bell, and we quickly tossed out a reward piece of food wrapped in confetti string that hit the water with a splash just to the left of the target. The shark quickly grabbed the food, cutting the string with his teeth. In this first critical test period, he proved he had associated pressing the target with getting his food by repeatedly pushing the empty target and then taking the food tossed to him.

The more timid female lemon shark took longer to respond to the empty target, but within three days both were working the target without hesitation. We had succeeded in "instrumentally conditioning" the lemon sharks. The nurse shark gradually gave up.

We came to know the sharks as individuals. Even though they looked identical from above (except for markers we put on their dorsal fins), it was easy to tell which was the male and which the female by the manner in which they swam, turned, and reacted to the target. The male was bolder, and often when he saw us bringing the target onto the platform, he would cut short his wide-circling swim around the pen, go to a position opposite the platform, and, as the target was being lowered into the water, swim directly at it and push it with force almost the instant it was underwater.

The female would hold back until the male had got several pieces of food and satisfied his initial hunger before she would alternate working the target with him. She would approach the target from her wide-swimming circle and often press it gently with the side of her head, just barely enough to ring the bell.

We started dropping the reward food farther and farther away from the target, giving the sharks only 10 seconds to get the food after the bell sounded. This was to test their ability to learn to make a correct turn.

For some odd reason, probably having nothing to do with Coriolis force (the effect of the earth's rotation which causes a deflection to the right in the Northern Hemisphere and to the left in the Southern Hemisphere),* the sharks in our pens almost always swam in clockwise circles. After pressing the target, the shark would have to make a counterclockwise turn to be in the direction where the reward food splashed down. They were slow to break their swimming habit.

Usually a shark would press the target, make a clockwise circle, and come back for the food; however, with a time limit of 10 seconds, as we moved the food farther away, they had less chance to reach the food on a clockwise turn than on a counterclockwise turn. After a few long clockwise turns and reaching the food just as the 10 seconds were up and the food was being pulled out of the water, the sharks learned to make the counterclockwise turn.

And they learned something else we didn't anticipate. It was a dramatic sight for visitors to watch us lower the target into the water, have the shark rush to push the target, make a left turn, then swim 8 feet down the side of the pen and catch the food dropped there as it hit the water. Of course, this is an easy trick to teach a porpoise or a seal, mammals with intelligence on the level of a dog; but for the lowly shark—which some taxonomists consider so beneath the bony fishes in evolutionary development that they won't even classify it as a fish—this was an accomplishment.

I decided to make it more dramatic by training them to take a fast swim way down to the end of the 70-foot pen for their reward food each time they pressed the target. It would also be a good

---

* I was somewhat startled when Dr. Lochner, an acoustician working with sharks in South Africa, came to visit the Lab and remarked that all the sharks (in his lab's tank) swim in counterclockwise direction. Just a coincidence, probably. The local sea-bottom topography and currents undoubtedly influence a shark's swimming pattern more than the earth's rotation.

movie sequence for the film record I was trying to make. But as the food was tossed out farther and farther away from the target, it was the female who caught on first that when the bell rang she could get the food faster than the male if she just kept circling the food area instead of going to press the target. When the male pressed the target, the female took his food. Both sharks quickly learned that the one who pressed the target had the least chance of getting the food. They started to hold back from pressing the target, and I had to move the feeding place back nearer the target to continue the experiments.

We made a crude test of an acoustical factor. The sound of the bell cued off the feeder that the shark had pushed the target and the reward food could be tossed out. We suspected that the sharks heard the bell and that its sound had reinforced their learning. What would they do if they pushed the target and the bell didn't ring? One day, when everything had been running smoothly for some days and the sharks were building up excellent scores in our record book, we didn't use the bell.

As usual, the male charged the target immediately and hit it. No bell sounded but we dropped the reward food. The shark turned counterclockwise but slowed down, and then, *instead of going to the feeding area, he returned to the target.* The next time he pushed the target, and for the rest of the test period, it didn't bother him that the bell didn't ring. He pushed the target and went for his food, which Sam Hinton says should be called the Nobell Prize.

We continued to run the tests for eight weeks until, just before Christmas, the water temperature dropped below 70°F and the sharks lost interest in eating and in pressing the target. Then we learned something about their winter "hibernation" and memory.

The water temperature in the pens didn't go up to the 70s again until near the end of February, and the sharks started showing some interest in food again. All during the ten cold weeks, we had tried feeding them every few days just with food on a string. On February 19, we put the target in the water and the two lemon sharks pressed it as though there had never been an interruption in their daily routine with the target. We continued feeding the sharks by making them ring the bell, preparing them for more complex tests.

Dr. Dugald Brown, Chairman of the Department of Biology at the University of Michigan, came to do some experiments at the Lab in late April. He was making some complicated physiological tests on isolated pieces of living tissue from a shark's heart, which he kept alive in a high-pressure tube in a saline solution bath. Sometimes he had to check this tissue at odd times. On the night of May 1 near midnight, Dr. Brown went to the Lab to check the temperature of the bath containing the muscle tissue. After that, he walked out on the dock with a floodlight to see what the lemon sharks were doing. They were copulating.

The copulation of large sharks has never been witnessed before or since, that I know of. Only small species of relatively slow-moving sharks (like the California horned shark or the European dogfish) have been seen in copula—the male partly coiled around the female as they mate while resting on the bottom of an aquarium.

A male shark can easily be told from a female, even in young embryos, by paired modified extensions of their pelvic fins. These sizable claspers are possessed by the males of sharks, rays, sawfishes, and skates, all of which have internal fertilization. During copulation, one of the two penis-like claspers is rotated into a forward position, inserted into the vagina of the female, and then the unusual head of the clasper is opened as a person might expand one's hand from a fist. Cartilaginous ridges and hook-like spurs form a complex pattern that differs and is characteristic for each species of shark. An experienced shark anatomist can often tell the species of shark by examining just a clasper. Similarly, Dr. Donn Rosen, the *gonopodium* expert on *xiphophorin* fishes, can tell the species of platyfishes and swordtails apart by just the structural pattern of the male's intromittent organ, part of the anal fin of these aquarium fishes.

Dr. Brown couldn't make out any details from the dock, but he drew a sketch of the lemon sharks in copula. "They kept right on swimming in wide circles around the pen. I watched them for one hour."

The sharks were mating side by side, heads slightly apart but the posterior half of their bodies in such close contact and the swimming movements so perfectly synchronized that they gave the appearance of a single individual, a two-headed monster.

The next night, I stayed up with cameras and floodlights ready for use, and for the next week I drafted all reliable friends I could to help take watches through the night, but the sharks did not copulate again. Once, at dusk, the male lemon shark swam unusually close to the female and then sank, frozen in a curled position, to the bottom of the shallow end of the pen for four minutes. The claspers were noticeably pulled to the left side and appeared enlarged and slightly pink. A remora that usually accompanied the female shark was attached to the clasper region of the male shark with its sucking disc. We wondered if some seminal secretion was being released by the male lemon shark and acting as an attractant.

The female paid no special attention to the male during his stops, but occasionally she would stop her swimming and rest alongside the male. Once they both stopped swimming and rested side by side for twenty minutes. Actually, it may be more work for a lemon shark to stop swimming than to swim in slow circles. While swimming, a lemon shark keeps its mouth slightly open, and water passes in and is flushed over the gills and out the five pairs of passive gill openings. The sphincter around the esophagus at the back of the shark's throat is held in a contracted position so water doesn't flow down into the stomach. The esophageal sphincter muscle probably gets tired and opens when a shark, worn out from fighting a hook and line, is towed behind a boat and water is forced down its throat. Very little muscular activity of the shark's body is needed to keep its streamlined body in motion. But when a shark stops swimming, it has to pump water over its gills to keep up an oxygen supply. The muscular apparatus to keep this pump system going and opening and closing the mouth and gill slits looks as if it uses more calories than are needed to keep up a leisurely swim.

All during May, the pair of lemon sharks seemed unusually close and, we imagined, affectionate. They often swam together, side by side or in tandem for long periods. Although they ate a little less food at this time, they worked the target very well, and we conducted many kinds of experiments with them, especially on their swimming pattern and ability to hear sounds.

Then unwittingly I made a decision with tragic consequences. I had read in books that since no cones had been found in histological preparations of the shark retina, sharks were color-blind. The retinal rods and a reflecting layer of guanin crystals at the back of a shark's eye were thought to be responsible for the shark's visual acuity in dim light and at night. And sharks seem more darkly poetic if you think of them as crepuscular creatures.

Just for fun, I decided to try a small test of our shark's color blindness. I painted the white target yellow. I figured the sharks would not notice the difference.

The male lemon shark lined himself up and rushed head long toward the yellow target as Oley lowered it into the water. Two feet from the target, he suddenly jammed on the brakes by lowering his pectoral fins, and did a back flip out of the water, sending a spray of water over Oley, Dr. Brown, and me. All three sharks in the pen began acting strangely, swimming erratically, fast, then slow, every which way, bumping into each other as they turned.

The male lemon shark never recovered from this experience. He refused food offered in any way, wouldn't go near even a white target again, and died three months later. His skin wrinkled as he lost weight. His swimming movements changed completely, his body twisted, and he kept his head turned to one side with his body muscles contracted on that side. He swam slowly, always at the surface, his back slightly arched and his tail almost half out of the water. The long upper lobe of his tail flapped on the surface of the water.

"Maybe he sprained his back on that back flip," Oley suggested. But then we noticed that on rare occasions the shark could straighten out.

We felt terrible about his death. For more than a year, he had been a part of our daily activities. We had examined hundreds of sharks by now, but it was very different to put the body of this lemon shark on the dock for a routine dissection. We did this after school hours because Tommy wanted to be there. The shark's liver was shriveled and leathery. We towed the remains of this once beautiful creature some miles out into the Gulf and watched it sink.

7   The Shark Hazard

AFTER OUR BIG male lemon shark died, we began making targets in different colors, shapes, and designs. We didn't spring any more surprises on our sharks but introduced them to changes gradually. The yellow target was not a shark repellent.

We trained other sharks to use it, and some learned faster and with more ease than with a white target. We learned that sharks don't like surprises. Some changes are O.K., such as the bell not ringing when the shark expects it. But when a shark is trained for a long time to a certain routine—perhaps over-trained—a sudden small change which would hardly bother a man in the same circumstances can startle a shark into erratic behavior: unusual swimming movements, change in swimming pattern, and disorientation, with the shark actually bumping into things it ordinarily avoided. Then the shark would avoid the target and stop eating for a day or a week or until we could coddle it back into its previous way of performing. Never since have we made any switch in our experiments with a trained shark which caused as dramatic an effect as the yellow target did on our big male lemon shark when he changed from an aggressive, confident, beautifully agile, and graceful creature to an uncoordinated and apparently neurotic shark with no interest in food. But we did wonder if some breakthrough in understanding the peculiar psychological condition that suddenly frightens a shark away and causes him to lose interest in food might give some clue in developing a shark repellent.

There is really no effective shark repellent. The Navy Shark Chaser, developed during World War II, is made up of several ingredients, the most significant being nigrosine dye. This dye (or almost any dark dye released into the water) will cause a shark to avoid the area or swim around it if it happens to be in his path—if there is no inducement to enter the dark area. However, if in the center of this chemical cloud (or in an area with any of the numerous other so-called "shark repellents" we've been asked to test) you place a piece of juicy cut fish, a hungry shark will swim directly into the "protected" area.

But the morale factor is important. After a shipwreck or other sea disaster causes vibrations that may attract sharks, a lone swimmer could panic and drown at the sight of sharks. He can be comforted by the thought that he has a little package on his belt labeled "Shark Chaser." And when he releases the contents of his package into the water and nearby sharks swim away, the improvement of his psychological state could be most important for his survival until help reaches him. Statistics of the infrequency of shark attacks and the fact that the sharks probably wouldn't bother him anyway are not much comfort when you find yourself in the water with them. (Your chances of being bitten by a shark when you're in the water with them are much less than your chances of being in a crash when you drive your car.) Somehow being mangled in an automobile doesn't arouse the same sense of horror as being bitten by a shark.

But after a sea disaster, sometimes a survivor in the water bleeds, vomits, has "nervous perspiration," or releases other excreta which might attract a shark. Dr. Albert Tester, who does research on olfactory and other chemical senses of sharks at the University of Hawaii's Marine Laboratory, visited the Lab and told us about his work. He found that almost all food substances he tried attracted sharks and produced the typical hunting response—surfacing, circling, rapid swimming, and, occasionally, jaw gaping. Oily fish such as tuna can excite a shark more than a "dry" snapper. Dr. Tester found that if he starved blacktip sharks, they responded to much smaller quantities of fish extract. His black- tip sharks showed an awareness of the presence of human urine in the water but were not attracted or repelled by it. Several species of sharks he tested showed an aversion to human sweat.

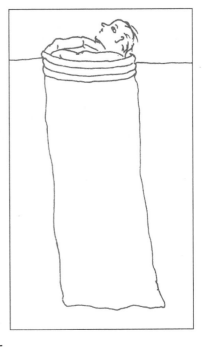

The plastic bag recently designed by Scott Johnson and tested during a six-month study by

Dr. Johnson at the Lab is probably the best protection against sharks for a man stranded at sea. This green bag which can be quickly unfolded and filled with water has an orange inflatable ring. It can be seen easily by rescue workers from a plane or boat, yet it screens the man inside from the shark that gets no olfactory cues. He sees only a large, inconspicuous bag, rather than dangling arms and legs.

Those of us who study sharks (or any animal, for that matter) know that the responses can be different depending on the species, the age of the shark, its state of health, its psychological state, its environment (water temperature and salinity), and its past experience. And, like people, sharks possess individual differences.

All careful investigators working with sharks are cautious to restrict their conclusions to the exact species and individual characteristics of sharks they work with and the exact conditions of the experiment. Sometimes scientific publications can be boring reading for a layman, as they seem filled with uninteresting details, but these details are very important when you are trying to fit together an accurate picture of a situation from which you want to draw true and meaningful conclusions. The Shark Research Panel of the American Institute of Biological Sciences keeps careful record of the details of all shark attacks on which it can get information, and every year reports and tries to analyze these data.

Something that may work as a repellent for one species of shark may attract another. This is true for some acoustical and electric "repellents." An expert diver I know claims he can repel sharks in the Red Sea by hitting two stones together; but in some little islands in the Pacific the natives attract sharks they want to catch by leaning over the side of their dugout canoe and hitting two stones together underwater. The species of sharks involved, however, are not known.

Some businessmen came to the Lab once to demonstrate an electrical repellent that they wished me to endorse. Engineers had worked it out in an electronics lab, and when they tested it on a shark, it had worked. They were vague about what kind of shark it was and under what conditions and for how long a period it repelled the shark. I put the gadget in our shark pen, and when we turned it on, the lemon sharks from the far side of the pen joined those nearby investigating the box discharging into the water.

Dr. Perry Gilbert's most dangerous incident in the water with a shark was when he was testing an electrical gadget "sure to repel a shark" and almost electrocuted himself.

There are supposed to be some effective electrical shark repellents being developed, but I have yet to hear about one that is safe for a diver to use, can drive away all kinds of dangerous sharks, and can continue to keep the shark away after it has discharged for some minutes and the shark becomes used to the unusual current.

Dr. Gilbert, who is head of the Shark Research Panel, has demonstrated that the "bubble curtain" (air pumped to an underwater pipe perforated with holes), thought to keep sharks away from swimmers, is not effective in deterring tiger sharks in captivity.

We all keep hoping that someone will develop an effective shark repellent, but after cooperating with dozens of people who claim they have one, and losing much time on inconclusive demonstrations, I no longer get excited when someone full of conviction and ready to take out a patent offers me the chance to test his shark repellent.

A species we have learned much about recently is the bull shark, *Carcharhinus leucas*. It is the most common of the sharks along the west coast of Florida, one we catch in quantity every month of the year. It can swim upstream from sea water into rivers and has been caught in the fresh water of Lake Okeechobee. In Central America, the bull shark has been called the Lake Nicaraguan shark. Because it lives in fresh water and has a somewhat stunted growth (reaching maturity at a smaller size than when in the sea), the Lake Nicaraguan shark was thought to be a separate species. It is greatly feared in Central America, where it is known to attack people.

Except for the east coast of Australia, the largest number of authenticated shark attacks in the world is in the Durban, South African area, where shark-watching towers are posted at crowded beaches so that swimmers can be warned out of the water when a shark is sighted and where all the doctors in the area are specially prepared to treat shark bites. The problem is so great and affects so many industries associated with tourism that much money is given to support shark research in South Africa. The expensive process of "meshing" for sharks is practiced around some public beaches, where the nets have to be cleared of sharks and mended every few days. The culprit mainly to blame is the Zambesi shark, which lives in the sea and goes up the Zambesi River. It is the same species of bull shark that is so common around Florida. Strangely, it is not known to have attacked a swimmer in this state except in the case of a "provoked" attack at the Miami Seaquarium, where the bull shark was in captivity and being handled.

The experienced skin diver will usually leave the water with as little commotion as possible if he sees a small six-foot white shark, tiger, or any so-called "carcharhinid" shark coming too close. But he won't mind a full-grown nurse shark of 10 feet, a basking shark of 40 feet, or a whale shark of 60 feet. I've never met the last two monsters when diving, but if I do, like most divers, I'll get into my boat as fast as possible for my underwater camera. When not photographing these big plankton-feeding sharks, divers like to try to ride them, and many divers have succeeded in hanging on to the tail or dorsal fin of a giant whale shark. As these sharks often bask at the surface, some fishermen—even little boys in the Bahamas—jump on the whale shark's back and run up and down until the shark, perhaps mildly annoyed, heads for deep water. The great size of these sharks could knock a ship on its side or injure a swimmer merely by bumping into him, but I don't know of anyone who has been hurt by a whale or basking shark.

The nurse shark can be a little dangerous if provoked. Its teeth are small but arranged in seven or more rows that can shred your skin. Some skin divers who have tried to ride a nurse shark by holding on to the tail or dorsal fin have a small scar from the experience. I've been slapped in the face by the tail of a nurse shark that looked dead as I bent over to examine it on our dock. I was thrown with a snap against sharp corals the first time I tried to ride a nurse shark by grabbing its tail. And I've had a long and wild ride and had my midriff and legs sandpapered when I once grabbed a nurse shark's pectoral fins and clung to its back while I was wearing a two-piece bathing suit. Several of us at the Lab regularly catch babies of this species by grabbing the tail with one hand and the back of the head with the other. I've never been bitten, but one of my friends has.

Jon Hamlin, the son of Alley Oop's creator, doesn't blame the nurse shark for his scar. Jon is a fine diver who has collected many small sharks by hand for the Lab, by pulling them out from under the ledges of Point of Rocks, Sarasota. Jon once grabbed a protruding

*Beryl is assisted in bringing in a tiger shark by Dr. Tony Perks,
an English scientist studying pituitary glands. C1959.*

tail belonging to a shark that was longer than he expected. He couldn't bear to let go of his prize and they tugged in opposite directions until the shark turned and grabbed Jon's leg like a bulldog. As soon as Jon released the shark's tail, however, it released Jon and swam straight out of sight.

Recording the measurements of a bull shark

For some reason, the nurse shark is not well known. Although it is common all around Florida and looks so different from other sharks, it is usually called a "sand shark," that catchall term applied to any of a dozen or more species even by fishermen. Or they may not recognize it as a shark at all. A male nurse shark made headlines in Florida newspapers until it was identified. "Mysterious monster, caught in St. Petersburg . . . 8 feet long, with whiskers like a catfish and a pair of legs without feet," read the account.

I wish I had a good story to tell about my most dangerous contact with a shark. Oh, I could probably embellish a few incidents when I was awfully close to some big sharks, but the incidents all had happy endings. I've been down in a one-man shark cage in the middle of the Red Sea with big offshore sharks swimming all around me.

121

Another time, I crouched in a depression in a coral reef as Claude Templier and I unexpectedly set off a feeding frenzy among seven sharks and hoped the air in our aqualungs would last until we found a relatively safe moment to swim for the surface. And once, at 80 feet underwater, I backed my air tanks against a vertical drop-off of over 4,000 feet as I watched Dr. Joseph François, just beside me, poke his shark billy at the head of a sizable shark that suddenly rushed toward us from the deep blue clean water near the Suakin Islands.

Just once my arm dripped blood after the teeth of a twelve foot tiger shark sank into it. I was driving my car down the Tamiami Trail when I stopped short for a red light. I was late on my way to give a lecture at Riverview High School, in Sarasota, and had some books and props on the seat beside me, including the dried and mounted jaws of a tiger shark we'd caught. As I slammed on the brakes, I stretched out my arm to try to hold the pile of material from falling off the seat. The students seemed to like the opening of my lecture that day, when I could demonstrate my fresh, though mild, wounds.

There have been only two bad shark attacks on the central west coast of Florida.* One happened in June, 1920, long before the Lab opened. I learned about it one day when I went to get gas for my car at the Texaco station in Englewood. The attendant looked hard at me and asked, "Are you the lady who studies sharks?"

He pulled up his pants leg and showed me a terrible scar. From the size of the single huge curve, I judged it must have been a shark well over ten feet, a tiger or possibly a white shark. The man,

---

* Editor's note: Statistics on shark attacks are maintained at the University of Florida Museum of Natural History (www.flmnh.ufl.edu/fish/sharks/statistics/GAttack/). From 1882 to 2008, 35 unprovoked attacks are listed for the Florida west coast; two were fatal.

Hayword Green, had been swimming off the Englewood Beach a good distance from shore beyond the sandbar, in water over his head, one evening at about 6:30, when he felt something grab his leg. He swam to shore and luckily was helped before he bled to death as many shark victims have.

A worse shark attack occurred off Longboat Key, Sarasota, on July 27, 1958, at about 4:10 P.M. The parents and brother of the boy, Douglas Lawton, who was attacked came to see me at the Lab the next day. "How could such a thing happen in such shallow water?" they asked. "What kind of a shark was it?" Doug was in Sarasota Memorial Hospital, where he had to have his leg amputated. He was eight years old. I went to see him and talked with the five witnesses: his parents; his aunt and uncle, who were on the beach; and his twelve-year-old brother, who was swimming with Doug and dragged him to shore with the shark still holding on to his leg. At the water's edge, Doug's uncle held the boy by the shoulders, his father grabbed the shark's tail and pulled, and Doug pushed the shark's head with his hand. The shark finally released its grip and floundered at the water's edge until it was in water deep enough to swim away. It was a small shark, about five feet long.

I went to the office of Dr. John Bracken, the pathologist, and he gave me Doug's amputated leg to examine. The shark had attacked the leg at least three times. The deepest bite, high on the thigh, made it necessary to amputate the entire leg. I found the clue to the species of shark in some superficial nicks made by the shark's teeth just below the knee. One tooth-mark, undoubtedly made by a front tooth, in particular showed that the largest cusp of the serrated edge was off center. Dr. Bracken took photographs.

Back at the Cape Haze Marine Lab I studied the photos, my

notes, and measurements. I got some clay and made tooth imprints on it, using preserved jaws of all the different species of sharks we had collected. The distance between the tooth-marks and the wide curve of the bite for a shark of five feet reduced the possible species involved down to four: the dusky, sandbar, bull, and tiger sharks. The place and time of year ruled out the first two species. The shape of the tooth-mark ruled out the bull shark. It must have been a young tiger shark. All the witnesses and Doug himself had stressed that the head of the shark was blunt, not pointed as they remembered pictures of sharks. The tiger shark has an exceptionally blunt head. They had told me the shark's body was streamlined, and Doug's father remembered that the base of the tail where he grabbed and pulled was narrow and felt sharp. The tiger shark has a small caudal peduncle with a keel on each side (the only shark with these keels that comes close to shore in the Sarasota area). The swiftest swimming sharks, especially deeper-water sharks like the mako, have these keels. Everything pointed to the tiger shark as the attacker except for the fact that none of the witnesses noticed any markings. But on looking over photos of tiger sharks of about five feet, I realized that unless you look at this shark from the side, the stripes are hardly noticeable. The witnesses saw the shark mostly from above. They knew it was a shark, and understandably they were so concerned about Doug that no one paid any attention to how the shark looked.

Why did the shark attack Doug? We took the *Dancer* to the spot where the attack had taken place, about 10 feet from the shore in murky water 3 feet deep. We found a long sandbar about 25 yards from shore running almost the entire length of Longboat Key, a distance of nearly eight miles. Low tide on July 27 was at 5:57 P.M.

We figured the depth of water over the sandbar at the time of the attack must have been only a few inches. The channel inside the bar was about 5 feet at its deepest point but led into deeper water (10 feet) in Longboat Pass at the north end of the Key and into New Pass (15 feet deep) on the south end of the Key. The water next to the beach dropped sharply from a few inches to 3 feet.

Possibly the shark had swum over the bar earlier in the day and then found itself trapped in the channel as the tide went out. Or it may have entered the channel from either of the passes at the ends of Longboat Key.

The victim and his brother were the only people in the water at the time, and the shark, swimming within the confines of this channel, might easily have detected the vibrations made by the boys slapping their flippers at the surface of the water, the irregular vibrations of boys playing rather than swimming. The victim's feet and ankles were not as deeply tanned as the rest of his legs, as he usually wore shoes and socks on his feet when playing in the sun. It seems possible that the shark, attracted by the vibrations made with the flippers, saw the pale lower portion of the boy's leg and struck at that point first, causing a large wound on the foot. The victim's left flipper was lost, presumably during the attack. The blood and the struggling victim could then have aroused the shark into a feeding frenzy, causing the repeated attacks.

All of us working with sharks in captivity get such a different attitude toward them that it is an effort for us to remember how dangerous they can be and to keep up necessary precautions. The vicious attack on Douglas Lawton by a young tiger shark seemed as incongruous to us as a dog lover might feel who keeps several boxers at home and hears about a mad dog attacking some innocent person.

Newborn tiger sharks are beautifully marked with black stripes and spots on their silvery white bodies. Hera and her friend Zen helped us deliver a litter of 37 tiger sharks, each one 2 1/2 feet long, from the uterus of a dying mother. (Tiger sharks are known to carry as many as 105 embryos at one time.) These babies were slightly premature but some started feeding and one became remarkably healthy. In three months it grew only two inches in length but doubled its weight. It fed from our hands and allowed itself to be stroked. If you walked to the side of its small pen, the baby shark would swim over to see you immediately. We had hopes of raising a tiger shark in captivity and finding out how fast this species grows and matures—something never done before. But this was cut short by a vandal who broke into the Laboratory grounds one night and beat the young tiger shark to death. We found the stout stick near the pen and marks on the bruised dead body of the young shark matched the shape of the end of the stick.

In the history of the Cape Haze Marine Lab, we had only one human mishap in the course of our daily work. In 1956, Beryl brought a class of school children onto the dock beside the shark pen to show them his favorite shark, Rosy. This big old nurse shark was at the bottom of the pen and couldn't be seen. Beryl, who often called her up in this way to feed her, splashed his hand on the water while kneeling on the feeding platform. He let his hand dangle over the side some distance above the water and turned for an instant to say something to the watchful children. Rosy lifted her head out of the water and touched Beryl's hand. It was such a smooth, gentle-looking maneuver that it could have been mistaken for a kiss. But the end of one of Beryl's fingers has been a quarter of an inch shorter

Dr. Breder (L) and Oley Farver, our second shark fisherman, observe the target training of lemon shark. Note the white square and colored (red) circle on the pilings.

ever since.

8 Scientists and Students

IN OCTOBER, 1958, in the midst of some of our shark-conditioning experiments, I took time off to give birth to Nikolas Masatomo Konstantinou and even up the sex ratio in my family. Dr. Finch, my obstetrician, had expressed some concern about my diving activities, which went on during the period of the shark-training experiments and beyond the time Niki theoretically was due. (Each test took less than half an hour each afternoon, leaving mornings free for diving.) Dr. Finch didn't object to my shallow-water dives, but diving at 70 feet, he warned me, might be a risk. He had never heard of any woman scuba diving in late pregnancy. And it wasn't until several years later, when I went diving with Japanese ama divers and could speak with them at length with the help of an interpreter, that I learned these women never stop diving because of pregnancy. They told me one woman in their group gave birth to a healthy baby in the boat returning from a day of diving to nearly a hundred feet, without using scuba. I'm sure my swimming activity throughout each pregnancy was a factor contributing to the extraordinary short and easy labor and birth of all four of my children.

I may have taken some risks, though. The pressure in deep diving may possibly affect an unborn child, especially during early pregnancy, when the embryo is still in the process of developing the main parts of its anatomy. Being pregnant for the fourth time in one's thirties can be uncomfortable. I was enormous by the eighth month. I had signs of varicose veins, and edema with swollen legs. But I felt completely comfortable when I was scuba diving, even if it did make me feel like a "Lady with a Sphere." If I could stay underwater for at least half an hour, all the weight and pressure was gone and my leg swelling and general edema disappeared. I made observations on *Serranus* living at the bottom in 70 feet of water shortly before Niki was born. It also felt good to dive 70 feet shortly after he was born. And before Tak and Niki were a week old, I put them in the sea. I let them dog-paddle at the surface, holding their chin, with occasional dunks completely in the water. At this age, babies in general have no fear of water and are practically natural swimmers. Even my daughters born up North went swimming with me before they were born. Hera swam in the YMCA pool in Buffalo, and Aya was the passive accomplice when I was spearing blackfish off Montauk Point.

Living and working beside and in the water somehow seemed to make childbearing and raising four small children easy to integrate with my job—especially with my parents nearby. The children often visited my mother and stepfather, who had their living quarters in back of Chidori, their Japanese restaurant. The children used chopsticks better than a knife and fork and were accustomed to Japanese food. They nibbled nori (seaweed) more than candy. I often left them for a visit with their grandparents during the day and picked them up on the way home from work, just before the restaurant got busy for supper. In our own house, Hera sometimes startled callers when she answered the door and announced, "We have some customers."

By 1959, many scientists were returning regularly to work at the Lab, staying for weeks or months at a time, often bringing their families along. Our special facilities for catching and working with sharks attracted numerous scientists to our laboratory. We were setting our lines almost continuously, sometimes nearly a hundred hooks at a time. Almost every day, we had at least a few sharks to dissect, and various parts of each shark were distributed among the scientists. Dr. John Heller, often with a team of scientists and technicians, and his wife and now four children worked at the Lab for one or more weeks at least once a year to collect liver samples. A Dr. John Phillips from England collected blood samples and extracted hormones. Dr. Anthony Perks and Dr. Hans Heller, also from the British Isles, competed for shark pituitary glands. Dr. Michio Oguri came from Japan for the bull shark's rectal glands. Drs. Wolfgang Klausewitz and Gerd von Warlert from Germany studied the locomotion and respiratory mechanisms of sharks.

Dr. John Heller and I had tried various ways to tranquilize our big sharks, so we could get in the water with them and handle them easily while attaching instruments and probes, without actually knocking them unconscious with MS-222. Our attempts failed, however. Once a healthy 12-foot shark recovered from the shock of being captured, started eating, and seemed to consider the Lab's pen his home, it was hard to catheterize it while it was conscious without an unwise calculated risk for the person holding the head end, even with the shark wrapped in a heavy tarred net.* Our injection of Thorazine in a shark recovering from capture acted as a stimulant and sent the shark suddenly rushing around the pen—and John, Oley, and me rushing out of it.

It was Dr. Fred Sudak, from the Albert Einstein College of Medicine, who found the simple solution. An injection of ethyl alcohol into the belly of a shark made it docile as a lamb in about twenty minutes and left the shark so good-natured for the next few hours (or so weak with a hangover) that we could handle even the largest sharks with ease, rolling them over, patting them on the belly or head while Dr. Sudak inserted catheters into the pericardial chamber and vessels and took electrocardiograms and blood pressure. Although a drunkard is not the best specimen for physiological tests, certain valuable information can be learned. Later, on an expedition in the Red Sea when we wanted to work with some large sharks in the water, I found that gin or vodka works too.

Parasitologists picked over the shark carcasses. Dr. Dorothy Saunders took blood to study microscopic blood parasites. Dr. Roger Cressi, from the Smithsonian Institution, went into all the crevices of skin, mouth, and gills for copepod parasites; and Dr. Lawrence

---

* Confiscated nets the State Board of Conservation had to take from fishermen who reduced the mesh size beyond the legal minimum by dipping their nets in tar for added strength. We applied to use these otherwise useless nets, which were taking up storage space at the Board of Conservation warehouse, for our work with sharks. We could seine sharks out of our pens and hold them relatively restricted in the nets.

Penner, from the University of Connecticut, continued to bring his wife, six children, and at least one or two of his graduate students every summer. He would set up his group in a disassembly line, assigning each assistant a different part of the shark's body, external and internal, for collecting every kind of parasite, from warty big green leeches to microscopic trematodes. One day I watched with awe the Penner operation, stretching from one end of the dock to the other. The line of assistants was longer than usual. Tommy Romans, who came to watch, had been given a pail of sea water and assigned the part in the lineup of rinsing out a shark stomach for worms. He loved it, of course. Tommy was probably the most promising young marine scientist in the local school with extracurricular activities like this to enhance his studies.

The Penner family is the most remarkably organized, energetic, yet relaxed family I've ever known. They often arrived at the Lab in their station wagon with a dead raccoon, skunk, opossum, snake, turtle, vulture, or any other animal that they had found on the road on the way over. These were dissected for parasites as they waited for sharks to be brought in from the line. Occasionally we had other big fish or a sea-turtle carcass for them. As my children grew older, they joined in the fun of days at the Lab when the Penner family came, along with any other children of visiting scientists, and drop-ins from the neighborhood. Some days we lost track of which children actually had permission to be at the Lab and which had crashed the party. Those who stood around with dry hands and just watched and made exclamations like "ecke" as two-foot-long worms were pulled out of shark intestines were checked and asked to leave if they had just strayed in. Those who dug in and helped without getting in anyone's way were allowed to stay and given jobs to do.

At noon, Mrs. Penner never counted heads as she passed out her endless supply of fresh carrot sticks, cheese, cold cuts, and homemade banana cake. No matter who or how many of any age were at the Cape Haze Marine Lab, when the Penners ate, anyone was welcome to join the meal. Sometimes this would include a group of school children and a teacher or more. Penner meals might include freshly caught conch, crabs, alligator steaks, turtleburgers, "pennerburgers," and an unwonted variety of wildlife you could be sure had been picked clean of parasites. Teachers such as grand old Bill Guild, who brought his gifted students from the St. Petersburg Science Center to the Lab, learned about the Penners and requested permission to visit on a day the Penner family was there. There were many such days of fun and learning when the Lab welcomed children. There were more and more visitors each year. Both Beryl and Oley liked to have children visit the Lab if they showed a serious interest in our work. An outsider casually dropping by the Lab on one of these busy days might find children everywhere— climbing around a mass of animal carcasses, clustered around Oley hoisting up a shark to be weighed, helping Dr. Penner pull a seine-load of fish onto the beach, looking in the microscopes, in the boat, lining the dock, looking at books in the library—children everywhere except in the shark pen. To such an outsider the Lab might look chaotic, as if we couldn't possibly be doing anything serious. Sometimes we didn't think so either, but years later, when some of these children came back to tell us proudly that they were preparing for college and careers in marine biology, we were sure we hadn't wasted any time on those days.

I found myself lecturing to groups of visiting school children several times a month. Beryl or Oley would take them on a tour outside the Lab first and show them the big shark pen, the smaller fish pens we had added on the other side of the dock, our tame pelican, the alligators born at the Lab from eggs we found in a nearby swamp, an outdoor cement pond with sea turtles we caught, and the ones we raised from eggs when Johnny and Barbara Bass collected mother and eggs from the beach at Placida. The eggs hatched 56 days later. The mother and her babies are the subjects of one of the most interesting movies we have made at the Lab.

When a first-grade teacher asked to bring a group of twenty-five students to visit the Lab, I hesitated. But after their orderly three-hour visit, the last hour of which we spent sitting on the grass for a question-and-answer period, I decided there was no need to put a minimum age limit on visiting classes. Their thank-you notes came in the form of charming colorful drawings and collages, depicting their impressions of the Lab, both real and imaginary. In one, I was being swallowed by a shark.

Dr. Penner was studying the life cycle of a parasite that has as one of its hosts the crowned conch, *Melongena*. From this he branched into studying the growth rate of the crowned conch. He placed tiny numbered metal tags on several hundred of the mollusks and released them in the bay next to the Lab, hoping to recover some the following summer.

That winter I was studying at my desk one afternoon when I looked out the Lab window and saw a man wading around the shallow water under our dock. When he came to shore, he assured me he wasn't bothering the sharks, he was only picking some shells. He showed me his bucketful. About a third had Dr. Penner's tags and the live animals still inside. The shell collector was surprised to learn that the people at the Lab "fooled around with anything except sharks."

We started getting letters from young students as far away as Alaska asking to be volunteer workers at the Lab for the summer. Some had well-to-do parents who would pay all expenses and who even offered to contribute money for our research work. Others asked for small financial help to pay their travel expenses. We couldn't accept them all. We received over eight hundred letters of this nature one spring when the Lab's summer program for young students was publicized.

We had to organize these summer programs with increasing care from the first year, when we accepted thirteen-year-old Carey Winfrey, to the summer ten years later when we conducted a program for eighteen high school students and teachers with the aid of grants from the National Science Foundation and the Selby Foundation. With this special financing, we could afford to choose the participants on the basis of merit alone and not have to consider the distance from which they came or their financial background. The foundations paid for their travel and living expenses if needed. That was the summer Father Robert Farmer, of the Blessed Sacrament Church in Tallahassee, and students helped me collect and study the precision data I needed to work out the tidal effect on the spawning time for *Serranus* living in New Pass. Father Farmer and two of the students who were trained in scuba diving worked out a method to record the changing speed of water flow along the bottom of the pass where *Serranus* mated. Father Farmer and I made records underwater on plastic pads, writing with ordinary pencils as we timed, to the second, the number of snaps and sex reversals and made a census of the number of mating pairs and trios in an area we marked off for study.

All the students had a chance to study the hermaphroditic fish, and squeeze out the eggs, and watch the development into larval fish. Several students stayed up all through the night to do this, especially Mike Hyson, who decided to make, as part of his project for the summer, a time-lapse movie, through the microscope, of the *Serranus* embryo's development from beginning to end. He was a careful, patient boy. I let him borrow the Lab's Bolex movie camera given to us by William Vanderbilt.

Mike set the camera over the microscope and made his time-lapse movie the hard way. He set the camera for single- frame exposure and snapped a picture every 30 seconds for what should have taken 18 hours. Another student slept on a cot beside Mike, who stayed beside the microscope except for breaks to eat or to go to the bathroom, and then Mike woke his friend to take over. The first night, the embryo was cooked to death by the heat from the microscope light. Then Mike devised a micro-depression slide setup where he could flush cool water over the embryos from time to time. He also switched the microscope lamp on only as he snapped a picture. To make life as comfortable as possible under the circumstances, for the developing embryo as well as for himself and his student friend, Mike moved his setup into the library wing, which we had air-conditioned to protect the books from mildew. I don't know whether the lower temperature of the room, the flashing light every 30 seconds, or the flush of cool water every 10 minutes did it, but the embryos took longer than 18 hours to hatch. Mike kept retaking movies, each time trying to correct the last of his faults. The final movie wasn't the best time-lapse photography I've seen, but it was good for a high school boy's first attempt.

One of our summer students decided to train newborn sharks for his project. Tim Wright from Ithaca, New York, and his young assistant, Bob Jackson from North Carolina, did a fine job and conditioned five baby sharks to press a target and ring a hell in half the time it had taken me to train the adults. Their final report was published in *Copeia,* a scientific journal for ichthyologists and herpetologists; I doubt if any of the professional scientists who read this journal could tell the authors of the paper were teenagers at the time of its publication. Both Tim and Bob are now studying for their Ph.D. degrees.

I wish I could say the same for Tommy Romans. Tommy's future looked so bright. But his love for diving and his lack of the rigid caution needed for it cut his career short in 1962. I was in Ethiopia, on my way to the Dahlak Archipelago on a six-week expedition, when I read a letter from my secretary at the Lab telling me that Tommy had drowned while trying to break a diving record. He could dive to 80 feet holding his breath, but he wanted to prove to himself and a group of boys that he could make it to 100 feet. He swam down a line to retrieve a marker placed at that level. He may have lost consciousness from hyperventilating* too much before the dive, or maybe his breath gave out at that depth where the lungs are so compressed that a diver loses his buoyancy and has to swim with effort to rise. No one was prepared to help him. It was over an hour before a scuba tank was brought to the scene and divers could recover his body from 140 feet below the surface.

The University of Miami has one of the largest centers for marine biological research in the world and is the center for tropical oceanographic research in the Atlantic and Caribbean waters. Since the first year I started work in Florida, I regularly visited this fine

---

* A series of deep breaths taken before a dive in order to lower the carbon dioxide content in the blood, which postpones the urge for breathing.

Institute of Marine Science to meet and discuss problems with fellow scientists who work there. Sometimes just two or three of us would meet in a small room crowded with jars of preserved fishes and talk over a common problem. Or we might gather to participate in a large formal convention opened by the most eloquent master of marine ceremonies, Dr. F. G. Walton Smith. He brought together marine scientists and the world's greatest fishermen at a common meeting ground where each could distill his experiences for the benefit of the other, talk in terms both could understand, and learn some of each other's ways. Some remarkable teamwork has come from these meetings and from the resulting cooperation of scientists and fishermen on life-history studies of fishes, especially from tagging. (My work with sharks always turned up something worth reporting at meetings, whether they were for such mixed groups or more limited groups of ichthyologists only.) Some of the marine scientists from the University of Miami came to work at the Cape Haze Marine Laboratory; three of them—Don de Sylva, Al Volpe, and Jack Randall—came especially to study and tag tarpon.

The children resented my frequent trips to the University of Miami for meetings, to use the library, to visit a new type of Lab, to witness some experiment in progress, or to examine some preserved fishes. "Are you going to Yourami again?" Tak would ask, and then beg to go along. So I started taking them when I could, combining my visits to the University with taking the children to the Miami Seaquarium, where we saw Craig Phillips cuddling a baby manatee he had just caught bare-handed and Captain Bill Gray bringing in some big sharks and feeding his famous white dolphin.

As a working mother, I have often felt guilty about the amount of time I spend at my work away from the children who have been cared for largely by my mother and governesses. So when I can, perhaps I go overboard to include them in my work and on trips. They have in some ways been spoiled or made blasé, and sometimes take for granted experiences that other children would consider fabulous.

In the summer of 1959, I got a phone call from the Art Linkletter program director in Hollywood, asking me to be on "House Party" to tell about my shark work. California had just had two widely publicized fatal shark attacks. It seemed an unlikely program for me to participate in, and I wasn't keen on taking the cross-country trip and losing at least two evenings with my family and two days' work at the Lab. But then they told me I could bring my children, and arrangements would be made for us to visit Marineland of the Pacific and Disneyland.

Niki was less than a year old and Tak was only two, both too young to take. But Hera, Aya, and their dolls took quite a wardrobe for what turned out to be a four-day trip with all expenses paid and was indeed "a party." We stayed at the Hollywood Roosevelt Hotel, and the first night we got there we had to see a Disney movie at Grauman's Chinese Theatre. We were up until nearly midnight matching our hands and feet with the imprints of movie stars.

I thought the girls would sleep late the next morning, but I was awakened before sunrise by the two of them in bathing suits. "Mommy, when are you going to wake up? Blondie and Dolly are ready for breakfast and we want to swim in the pool." They dragged me out our sliding glass wall that opened into a patio and pool.

At the studio they were fascinated, as I was, by the professional makeup job, all the paints and powder put on my face by a long, thin-fingered man who tried to talk me into letting him tweeze my eyebrows. He gave up when the children told him about my work, cutting up sharks. The girls watched the program on a series of six monitors in the control room, where the director and technicians patiently put up with all the questions. We were picked up and chauffeured wherever we wanted to go. After the program, we went to Disneyland and the children were allowed to go on all the rides and eat anything they wished. I was afraid they might get sick, but they weren't even tired at the end of what was, for me, a very long day.

The next morning we had to get up early for the long drive to see Marineland. In order to get a little more rest for myself, and before the girls decided to go for another sunrise splash in the pool, I told them we were going to have breakfast in bed and they could order anything they wished.

"Steak!" they screamed, "Mine rare," Hera demanded, rubbing her tummy and rolling her eyes. "My dolly wants cream cheese," Aya said, translating a mechanical squeal.

The steak did sound sensible after all the cotton candy, popcorn, hot dogs, and finally the rich Chinese dinner we had had the day before.

I phoned room service and, yes, they could send up three steaks.

"Don't forget the cream cheese for my dolly."

"Oh, Mommy, puleeze can I have a chocolate sundae for dessert?"

"My dolly and I want strawberry."

The voice on the phone sounded as if it were taking the most ordinary breakfast order and, in answer to my apologetic question, said of course they could make sundaes at 7:00 A.M. The waiter who wheeled the breakfast on a drop leaf table into our room took a long look at us and asked, before setting it up, if we had ordered steaks. It still bothers my conscience when I remember signing the check for $28.00. The girls pointed out to him that he hadn't brought enough silverware but it was O.K., as their dollies could use the same forks they did.

The steaks were enormous and we couldn't finish them. I made steak sandwiches to take with us and used up the rest of a large basket of delicious rolls and a platter of butter pats to do so. There were only a few crumbs left on the table when the waiter came back to remove it.

At Marineland, Curator Ken Norris gave us a personal tour, and the girls and I were allowed to go out on the feeding pulpit over the huge tank; the girls clung to my skirt as Bubbles, the whale, leaped out of the water and placed a giant rubber dumbbell in my hand.

I don't know if any of my children will become marine biologists, but I think they will appreciate the sea, and their exposure to marine life may help them choose more wisely what field they want to work in someday. There are so many areas related to the sea opening up for young people who wish to work as physicists, chemists, biologists, engineers, doctors, technicians—whether it be as a conservationist trying to increase our dwindling whale populations; an aquaculturist growing new food fish; a mathematician programming computers to analyze underwater sound recordings or the vocal communications between porpoises; an architect designing an underwater city; a frogman; a physiologist or a psychiatrist who studies the effects of deep diving on the body and mind of man; a policeman of an underwater park; a cook on a submarine; a plumber for underwater housing; an attorney specializing in international laws to protect marine life on the high seas and on our continental shelves from overfishing;

pollution experts—the possibilities are almost limitless.

Perhaps Hera's present absorption in writing poetry and illustrating her thoughts in Beardsley-like pen-and-ink sketches and abstract clay sculptures may someday combine with her love for skin diving and watching sea life to produce some combination of art and science that will enrich her life. Aya and Tak, with their swim- and gymnastics-team activities and tennis-tournament participations, seem too energetic at present to settle down to any serious study. Niki will be a philosopher of some kind. They all love the sea and its life. That's enough satisfaction for me even if their future careers have nothing to do with oceanography.

9 *Mesoplodon*

EXCLUDING SHARKS and manta rays, a fish over six feet long is something you don't come across often, even if you're a fisherman or an ichthyologist. After several years of collecting and skin diving along the west coast of Florida, I had never seen a tarpon, a giant grouper, or a barracuda that actually measured over six feet. So I got quite excited one morning in April, 1959, when I got a phone call from a commercial fisherman who identified himself as Captain Bruce Ellison, owner of the Kozy Kitchen at Boca Grande, and told me, "There's a fish washed ashore here that's about 15 feet long. I thought you might like to take a look at it."

At first I thought it must be a shark, but Captain Ellison assured me it was not. The vultures had already eaten away most of the flesh and exposed a backbone, long ribs, and bony skull, so it couldn't be a shark. Captain Ellison was also sure the skeleton was not that of a porpoise. It certainly sounded worth investigating. Oley Farver, Dr. Dugald Brown, and I drove to Boca Grande to meet Captain Ellison.

On the way over, I tried to figure out what kind of skeleton it might be, An oarfish was a possibility that came to mind. The oarfishes are a little-known group of deep-sea fish. On very rare occasions, a specimen is collected when a stray, probably sick, individual gets caught in shallow water and is washed ashore. One was washed ashore on the west coast of Florida in 1954 near Clearwater. It was ten feet long and showed the typical characteristics of the family of oarfishes. It was long and eel-like but flattened side to side; its first dorsal spines were far forward on the head and elongated, and they formed a crest of crimson that looked like a tiara. It was brought to Captain and Mrs. Barnett Harris, who were preparing a museum for the city of Clearwater. It was the rare *Regalecus glisne*, the first record of a specimen from the Gulf of Mexico. Could Captain Ellison's fish be the second record of this oarfish?

Captain Ellison was waiting for us at the Kozy Kitchen. "I've got the skull in the back of my car," he told me. He opened the trunk of his new Oldsmobile. A skull was lying on some newspapers; although most of the flesh had been cleaned away by the vultures, there were still some pieces of sand, seaweed, and decaying flesh clinging to it, and maggots were crawling in and out. It was too massive a skull to belong to the delicate oarfish. It was a cetacean skull.

My first reaction was disappointment because the skull was not that of a fish. Then I realized that it was too large to be the common bottle-nose porpoise, and there was a peculiar curve to the upper profile of the skull that wasn't like a porpoise skull. We drove over the railroad track and walked across to an isolated beach to see the rest of the remains, a typical cetacean skeleton. The long ribs were still attached to the breastbone, although the whole rib cage was distorted, squashed to the side, and disconnected from the backbone. The shortened upper arm bone and the lower arm bones were lying nearly in place and still attached to the shoulder bone. The small wrist and "finger" bones that hide in the paddle-like flippers of cetaceans were missing, probably scattered or carried off by the vultures. The long backbone was intact and ended in a fluke, the horizontal tail of whales and porpoises, which is a good way to distinguish cetaceans from large fishes. The fluke had some meat and skin still clinging to it, and it looked as if the cetacean had been black with a pale border on the end of its fluke. It measured 12 feet 9 inches long. From the size alone, it was too big to be the common porpoise.

We took the bones back to the Laboratory, and I got out our books and papers on cetaceans and found, mostly from comparing the skull, that we had the skeleton of one of the little-known whales commonly referred to as "beaked whales," because of the long beaklike protuberance of the jaws similar to those of the bottle-nose dolphin. It was a whale of the genus *Mesoplodon*, whales that possess an unusual pair of enlarged teeth in the lower jaws. Unlike other whales and porpoises, *Mesoplodon* have neither baleen hanging from their jaws nor rows of teeth, only a pair of lower-jaw canines which in some get to be so long that they break through the top of the head and stick out like a pair of horns when the whale's mouth is closed.

In order to tell what species of *Mesoplodon* we had, we needed to have the lower jaw, which we had not found. Oley volunteered to go back and give the place a thorough search. He took a shovel and a wire-mesh box and dug through the sand and sifted it, hoping to find fragments of the lower jaw or the two big missing teeth. He finally waded out into the water, searching along the bottom, and about thirty feet from shore he recovered most of the lower jaw, broken into pieces. He was sure he couldn't have overlooked the teeth if they were anywhere around the place where the whale was found. We wondered how the lower jaw got so far away from the rest of the skeleton.

We pieced the lower jaw together. The empty sockets of the two large teeth were two inches wide and an inch and a half deep. We would have given anything to see the ivory-like teeth that belonged in those sockets. The sockets were three inches from the tip of the lower jaw on the joining portion of the left and right lower jawbones, which then spread widely apart. The position of the teeth sockets was very puzzling, and the more I read about the few and rare species of *Mesoplodon*, the less ours seemed to fit any of the known species. I rushed a letter off to Dr. Robert Cushman Murphy, one of the members of the Laboratory's Board of Advisors. I sent Dr. Murphy a sketch of the lower jawbones and asked if he could help identify the whale and advise me what to do with the skeleton. He turned my letter and sketch over to Dr. Joseph Moore, of the Department of Mammals at the American Museum of Natural History, the world authority on *Mesoplodon*.

Dr. Moore was extremely pleased with the report. From my sketch of the lower jaws, he believed the specimen was *Mesoplodon gervaisi*. The first specimen of this species ever known was a dead male found floating in the English Channel more than a hundred years before. Since then, ten females have been seen. The Boca Grande specimen was the second male specimen known. The sex could be told from the large size of the tooth sockets; the teeth of females are small. The Boca Grande whale was also the first record of this species from the Gulf of Mexico.

We cleaned the skeleton and were very proud of our rare specimen. It was, of course, the largest in our growing collection of marine plants and creatures from the west coast of Florida; it took up a lot of space and really belonged in a museum. So we decided to ship it to Dr. Moore at the American Museum of Natural History instead of making Dr. Moore come to the Lab to study it. But Dr. Moore wasn't satisfied. "If only we could have the teeth, too!" he wrote back. The teeth were extremely important for his studies, and he suggested that since they must be objects of rare beauty—a flattened ivory—they might have been taken as souvenirs by someone who saw the whale before Captain Ellison found it.

By now, word of the rare whale had reached the newspapers, and various reporters came to interview us and to photograph the skeleton. Captain Ellison was credited as being the first to see the whale and report his find to the Laboratory. It was a good opportunity to stress the importance of the missing pair of teeth.

One of the responses to a newspaper article on the find was from five men who claimed they saw the whole whale before Captain Ellison, and demanded that they receive credit with their names in print. Therefore, I would like to mention that A. W. Landers, W. L. Moore, L. E. Brown, R. L. Sands, and E. L. Pepples saw the whale before Captain Ellison and did not report it.

We did hear from a family who subsequently earned the gratitude of scientists and the admiration of all who know their role in the *Mesoplodon* find.

Captain Claude McCall came to see me. He was a commercial fisherman and worked hard to earn his living. While looking for new fishing grounds, he had passed the isolated beach near Boca Grande and spotted the whale. He saw it before Captain Ellison did and, noticing the unusual teeth, took them to his ten-year-old son Terry. His son was a budding naturalist and collected all kinds of specimens. His mother called his room, which was full of glass jars and boxes, Terry's Junkorama, Captain McCall told me. The pair of *Mesoplodon* teeth was now in that room and was Terry's prized possession. Captain McCall looked around the Laboratory.

"My boy would sure like to see this place," he said. "I don't think he'd want to give up the teeth, though."

I hurriedly wrote Dr. Moore about the situation, and he wrote back authorizing me to offer any price within reason for one of the teeth.

When the McCall family came to the Lab to leave Terry with us for the day, Terry had brought the teeth with him. I'd told his father I wanted at least to see the teeth and have them photographed. The teeth were truly beautiful and a rare collector's item. Terry was obviously very proud to be the owner of ivory of such value and importance.

Before bringing up the subject with Terry, I called his mother aside and told her about the offer from Dr. Moore.

"Claude and I have already talked about this," she said.

"We thought you might want to buy the teeth, but we don't want Terry to sell them. They mean too much to him." I dropped the subject of buying the teeth.

Terry was a delightful boy with the wonderful uninhibited exuberance and curiosity about nature, especially the sea, that children naturally possess. I knew his father must have told him many tales of fishing adventures and passed on a father's love for the sea. Terry, who enjoyed studying all his "junk," was pleased to know that such a hobby could also be a serious profession in which you could continue the fun under the label of scientist, if you were willing to put in extra years of school and study. I showed him through the Laboratory and told him about the fish in our aquariums, and the experiments and studies we were making.

When his parents came to pick him up at the end of the day, he wanted to know if he could come back to visit the Laboratory again. The three McCalls went into a huddle and spoke quietly for a short time before they came over to say goodbye. They thanked me for letting Terry spend the day at the Lab. As they were leaving, Terry reached in his pocket and pulled out the teeth that we had photographed from every angle that day.

"I'd like to give these to science," he said as he placed them in my hand, and the McCall family left as I fumbled for some words of thanks and goodbye.

Dr. Moore was also deeply touched when I told him, in a letter accompanying the teeth, how the boy gave them up. He sent Terry a grandiose official letter of thanks from the Museum, and eventually, when Dr. Moore's study of the whale was published, he gave one of the kindest acknowledgments I've ever seen in a scientific paper. He expressed his gratitude to all of us and to Terry, who "at the age of ten" was able "to gain for himself some grasp of the character of scientific work so that he voluntarily presented the teeth of this whale to the Museum."

Dr. Moore also wrote and asked me to find out if there was anything the Museum might have to spare that Terry would like, so that the Museum could send Terry a gift in exchange. The next time Terry visited the Lab, I put the question to him directly, expecting he would then take a great deal of time to think over the possibilities. He answered me without a moment's hesitation: "I've always wanted a grizzly bearskin." Terry didn't have to wait long before he nailed the skin of a grizzly bear from Dr. Moore on the wall of his treasured Junkorama room.

Since my experience with Terry and the teeth, I've realized the importance of taking precise information and records on whales. These large creatures, when washed ashore dead, are almost always noticed by the local people, but unless it is in a populated area where the stench creates a situation resulting in publicity that reaches the attention of a whale specialist, valuable remains may be lost to science.

I have investigated reports or come across remains of only four other whales since that time. In 1962, several people in Sarasota reported seeing a pair of large whales from their boats. I sent Dr. Moore a detailed report, based on interviews with the witnesses (one was a minister) and one fuzzy photograph. This turned out to be the first record of right whales in the Gulf of Mexico. In June, 1964, after attending the Fifth Annual Meeting of the Underwater Society of America in Mexico City, Don Pablo Bush Romero took some of us on a trip along the eastern coast of the Yucatán Peninsula. We had been diving on the wreck of the *Matanceros*, which sank off the coast of Quintana Roo in 1742. I was still enthralled with some beautiful crucifixes that I'd found in the coral-encrusted wreck that day when a group of us strolled along the beach in the evening and stumbled across the skeleton of a whale. It was a male *Mesoplodon mirus*, the only record of this species from Mexico.

The other two whales I investigated were common whales, but tracking down the corpses was fun and adventure that one remembers long after forgetting the stench of the actual situations. We managed to save the body of a pilot whale from cremation in the garbage dump at Venice, Florida, by enlisting the help of the Highway Patrol. It had been taken to the dump for burning before anyone thought to try to identify it. The dissection of even this common whale was fascinating. I wondered what kind of headache the poor creature had suffered as we removed hundreds of nematode worms, each about two inches long and still wriggling, from the head cavities.

Our expedition to examine a truly gigantic whale in the summer of 1963, weeks after its body washed ashore near Lower Matecumbe in the Florida Keys, was one our group from the Laboratory will long remember. Many baths have since removed the strange odor that clung to our skin after we waded and slid around waistdeep in a mixture of bay water and rotting flesh and blubber. I seriously considered wearing scuba gear even though my head was not underwater.

The beautiful 900-pound skull of this sperm whale, cleaned and mounted, now decorates the entranceway of the Laboratory. Those of us who gathered up the parts and brought them back complained at the time, but later we all agreed that if you can stand the smell, dead-whale hunting is a rewarding experience.

I was appointed by the Committee on Marine Mammals of the American Society of Mammalogists as their official representative to record stranding and sightings of cetaceans in the southeast sector of the United States; the long Florida coastline makes this the choicest sector in the country. It was an honor, of course, for an ichthyologist to be appointed a whale watcher. I never thought when I started studying little guppies as a child that my zoological interest would grow so big.

10　Sinkholes and "Rapture of the Deep"

I SUPPOSE IT WAS INEVITABLE that with so many fresh-water springs not far from Charlotte Harbor, I would dive in Florida's fresh water as well as its famous salt water.

My first dive, shortly after the Lab opened in 1955, had been in the cold clear water at Silver Springs. I was invited to have an underwater look at a fresh-water porpoise from South America, the first ever captured and exhibited in North America. I was also invited to tell about our new laboratory on a radio program conducted by a diving disc jockey who greeted me with a towel when I swam down to the bottom of Silver Springs and into the big air bubble in which his studio was set up.

Florida has many water springs. Certain species of marine fishes get into these springs when high tides or rains connect them with the adjacent sea. Mullet and tarpon are common marine fishes that adapt to and live in the springs. Some springs are famous for their underwater scenery and exceptional clearness. Silver Springs, Wakulla Springs, Weeki Wachee, and Rainbow Springs are attractions where tourists can look at the fish through glass-bottom boats and sit in an underwater theater and watch "mermaids" perform a ballet.

Unlike most Florida springs, Warm Mineral Springs and Little Salt Springs, located in the outskirts of Venice not far from the Laboratory, are not clear. It was in these two springs that I made my weirdest and most dangerous dives—into waters where I couldn't read the depth gauge on my wrist with a flashlight directly over it, and to depths where nitrogen narcosis hits you unexpectedly. We had to feel around blindly for the human remains we sought of people whose bodies got there many years before we did. And even though there are no sharks around, the simple wet darkness can be terrifying.

My diving in these two sinkholes started in 1956 because of a letter I received from Dr. Albert Eide Parr, who was then the Director of the American Museum of Natural History in New York. He needed a specimen of a small tarpon and asked if I could send him one. It would have been no problem to send him a 4- or 5-foot tarpon; sports fishermen were catching them daily around Boca Grande. But one "under 2 feet" was not easy to find. Beryl told me that he had caught small tarpon inland, in brackish and fresh-water ponds and ditches, with a fly rod.

One year they might be plentiful; the next few years, in the same spots, you couldn't find one.

"There's only one place nearby I know where there's a couple of small tarpon right now," he said. "We could get one in a seine, but we'd have to get permission from the owner first."

Sam Herron, Manager of Warm Mineral Springs, gave us permission to dive and seine there. I looked forward to diving in Florida's only warm spring. People come to drink and bathe in these hard and sulfurous waters, which have a temperature that is always higher than 80°F. They also take mud baths at the shore of what is called in advertisements "The Fountain of Youth." It was Warm Mineral Springs that Ponce de León supposedly heard of from the Indians, who knew about its curative powers.

Iris Woolcock was such a good example of this "cure" that Sam Herron had hired her to work for the attraction and encouraged her to tell everyone how she came to Warm Mineral Springs crippled with arthritis and other ailments and became the picture of health after a few months of daily bathing in the springs.

Iris had also taken up skin diving and offered to be our guide. She was well into her sixties, but she looked marvelous in a bathing suit and could have passed for forty. She went snorkeling in the springs every day and had become fascinated with the melanistic population of mosquito fish (*Gambusia affinis*); the number of killifish (*Cyprinodon variegatus*) that had developed tumors; the strange slimy vegetation that, when the strong sunlight shone, let off bubbles which, as they escaped from the anchored plants, took part of the green-black slime along and looked like shiny comets with tails, rising toward the surface.

Of course Iris knew the habits of the two tarpon. "They're blind, you know, but can feel your swimming movements. If you float up to them quietly, you can sometimes touch them."

I brought Hera and her friend Zen along. They helped Beryl get the seine ready while Iris and I looked around with our face masks. We had to wade through soft mud, and as we entered the water every footstep stirred up a cloud of muck which cut down to almost nothing the already poor visibility in the milky green water. "Early in the morning, when the sun is low, you can see much better," Iris told us.

We swam around the circular spring. A thin layer of the blackish slime covered everything except the hundreds of tiny mosquito fish and killifish that swam near the shore. As we swam toward the center of the spring, the bottom sloped steeply at 20 feet, turned back into a cave-like ledge, and then, at about 40 feet from the surface, dropped down into a huge black hole. It was a spooky place.

Then we saw a tarpon. It was a small one about two feet long. Its eyes were bulging out of its head and covered with cataracts and a web of bright red enlarged blood vessels. The water may be good for arthritis, but I wouldn't like to trust my eyes in it too long.

Our helpers, Hera and Zen, were having fun sinking their feet into the warm soft mud, wiggling their toes, and stirring the bottom all along the shore. Beryl and I had to pull the seine through the blackened water a number of times before we got the tarpon that Iris kept swimming after and heading toward our seine. The tarpon looked old, probably dwarfed, and was gray rather than the shiny silver of tarpon caught at sea. But it was the best small specimen we could find to send to Dr. Parr.

On weekends, I went back to Warm Mineral Springs to dive around the eerie edge with Iris. I took the children, and we learned to ease into the warm water without stirring the mud too much. We had picnic lunches on the grass near the mud bathers, and the children watched wide-eyed as clients smeared black mud over their faces and corpulent bodies. Of course, the children had to imitate this, and added their version of a mud shampoo. It kept them amused as Iris and I explored the slimy sloping bottom—on each dive going a little deeper toward the large central sinkhole.

I began trying to identify the organisms that lived in Warm Mineral Springs. The greenish-black slime was made up of microscopic algae, mostly of the blue-green type. I couldn't identify for sure a large plant that grew to a height of several feet and the snail that crawled on it. So Oley and I packed a sample of the plant in one box, and a group of large snails in another box.

We sent one box to Dr. Harold Humm, the authority on aquatic plants, at Duke University, and the other box to Dr. William Clench at Harvard University, who is a noted malacologist. I wrote separately to the two scientists, asking Dr. Humm to identify the plant and Dr. Clench to identify the snail. I assured Oley these experts would send back identifications shortly and told him how cooperative scientists were about such matters.

A few days after we had mailed the samples and letters, Oley happened to see the carbon copies of the letters I wrote. "Oh-oh! I guess I didn't hear you right, Genie. I think I sent Dr. Humm the snails and Dr. Clench the plant!"

I hurried off letters of explanation but too late. They crossed letters in the mail from Dr. Humm and Dr. Clench. Dr. Humm identified the plant as *Entophysalis rivularis* (Kutzing) Dronet, and Dr. Clench identified the snail as *Amnicola augustina Pilsbry*. And they added some comments in their letters. Dr. Humm reported that he had a little difficulty finding the microscopic alga growing on the snail's shell. Dr. Clench wondered why I had included so much plant matter. It had taken him quite a bit of time locating the young little snail in it!

Iris and I finally put on scuba gear and ventured down the vertical drop-off into the sinkhole. None of the fish went below 30 feet. Below that depth, there was no sign of any animal life and the black slime everywhere absorbed the meager light from the surface. In some places, we found that by peeling off this velvet-like black skin, we could uncover a pale sandy layer that reflected some light and brightened the spooky caves we swam into.

Iris had told me all the myths she had heard about the spring. About divers trying to find bottom, who never returned. And about one diver who found a horse and carriage on a deep ledge. I wondered what might be on the floor of the cave-like ledges we saw, under the black slime. But in the dim light I didn't feel like groping around.

Iris told me about Little Salt Springs. "Hardly anyone ever goes there—it's way back in the woods, hard to reach, and the water is cold." It was Bill Royal who talked me into diving in Little Salt Springs.

Bill Royal was a retired Lieutenant Colonel from the Air Force, a grandfather in his mid-fifties. We had dived together many times in the Gulf of Mexico, where he helped me collect specimens of marine life and showed me where to find fossil mammoth and mastodon teeth off the Venice Beach. He had started the Venice Dolphin Divers Club and was its first president. He was an outstandingly superior diver and made the rest of us feel like beginners.

He had dived with me along the beach at Longboat Key where Douglas Lawton had been attacked by a shark the day before. I had wanted to see if there was any peculiar condition in the shoreline that might account in part for the attack. Bill Royal had a large knife and was hoping we'd see the shark. If we had, I'm sure Bill would have tried to kill it with no hesitation. His hobby, when he was stationed in the Pacific during World War II, was riding and killing sharks he met underwater.

Bill told me he did not succumb to nitrogen narcosis or what Captain Jacques-Yves Cousteau has poetically called "rapture of the deep." I tried to figure out why, and discussed it with a doctor friend who was an anesthetist. He told me that every once in a while doctors come across a person who is very difficult to anesthetize. The next time I saw Bill, I asked him if he was ever anesthetized in a hospital, and he said, "Oh, yes, once, and they had a helluva time putting me to sleep."

I had never seen Bill cold, tired, or hungry and I had been with him on trips where good divers less than half his age were exhausted, goose-pimpled, and anxious to go home. Once we were all complaining to Bill that we were hungry and in our little boat couldn't cook the fish we had speared. I'm very fond of eating raw fish, and the other divers were so hungry I convinced most of them it would taste good. But we worried about parasites. The fish we had speared were bottom fish that could have been in shallow water close to shore and picked up parasites dangerous to man. Bill solved that problem by promptly spearing a good-sized Spanish mackerel, a fast-swimming surface fish that's delicious and safe to eat raw. I filleted it and passed the pieces around. Everyone who tried it liked it. Bill stayed in the water, and when I asked him if he was afraid to come aboard and try some, he answered, "Oh, I've eaten lots of raw fish in Japan. Throw me some." Then he put on an act, barking like a seal and swimming up to the boat with almost the ease of the animal he imitated. We tossed him pieces of raw fish and he caught them in his mouth. Bill continued diving and came up with a slimy purple-black blob about four inches long, which he tossed into the boat. It moved when we touched it. Bill said he had seen it crawling along the sea floor and wondered what it was. All the others wanted it thrown out of the boat, but none of them would pick it up. It was a type of sea hare but one I hadn't seen before. When I got it back to the Lab and some books, I found out it was *Aplysia floridensis*, one of the rarest mollusks known; only a few had been collected before.

But Bill knew he had made a really rare discovery when he dived in Little Salt Springs. He'd been trying to get me to dive there with him for some time. "No one has ever dived in this spring before. There's no telling what we'll find. Maybe it's the deepest spring in Florida." Bill made the first scuba dive into Little Salt Springs in January, 1959. He went down 70 feet and found a cave containing stalactites. A young woman anthropologist, Dr. Luana Pettay, happened to be visiting her parents in Venice at the time. She was not particularly interested in skin diving, but she was interested in anything that might lead to some evidence of the way of life of the prehistoric Indians in Florida, and she went along and encouraged Bill to explore the springs. She was jubilant in March when Bill dived in the springs for the second time and found a fossilized human humerus on a ledge above the cave.

Bill had been excited from the time he found the first stalactite. He got in touch with Dr. H. K. Brooks, a skin-diving geology professor at the University of Florida. According to Dr. Brooks, the water level in Florida hasn't been low enough for stalactites to have formed in a cave 70 feet below sea level for at least 6,000 years. Bill called me up and said, "You've just gotta come along and see the cave." He had arranged a diving trip with Dr. Brooks and Professor

Lackey, a microbiologist from the University of Florida. Another diving friend of ours, Bill Stephens, a fine underwater photographer and the editor of *Florida Outdoors* magazine, also wanted to join the trip. His wife, Peggy, offered to babysit with their four and my four kids while we dived.

It's a good thing we didn't bring children. There was quite a group of us when Little Salt Springs was explored by divers for the third time, on March 9. Bill Royal was the modest leader of our group. Iris Woolcock joined, too. At 7:00 A.M., we met at her house near Warm Mineral Springs for hot coffee before we started off. Professor Lackey planned to collect his samples at the surface. The rest of us were prospective divers: the two Bills, Iris, Dr. Pettay, Dr. Brooks, Dr. Marwin (a dentist), and two Venice skin divers, Norman Rack and Bob Chapman.

We had to approach Little Salt Springs from State Highway 41 at a dirt-road entrance about a mile south from the main entrance to Warm Mineral Springs. We got the key to a wooden gate after we signed a paper stating that the Mackle Company, which owned the land, would not be held responsible for anything that happened to us. After going a mile on the crude dirt road, the cars were getting stuck in the mud. Iris and I abandoned my Volkswagen and joined Bill Stephens in his station wagon. We were all covered with mud from pushing the cars when we arrived at the spot nearest the springs where cars could go. Then we had to continue on foot for another fifty yards across softer mud. There were some wooden planks Bill Royal had previously put there, but they were slippery and we had quite a time carrying the air tanks on our backs and other gear in our arms while walking across the planks to get to the springs. Dr. Marwin slipped off onto a broken bottle. We all slipped and sank up to our knees in mud. At the rim of the springs, there was a high spot of dry land where we could get our diving gear assembled.

Little Salt Springs is about 250 feet in diameter. From the rim, the muddy bottom slopes gradually down to a depth of about 60 feet; then there is a vertical drop into a deep central hole about 70 feet in diameter.

Bill Royal dived alone into the dark central hole with an underwater light. The rest of us explored along the slope. Dr. Marwin and I buddy-dived and, at about 50 feet, circled the spring. When using scuba, you should always dive with a partner and keep each other in sight in case of trouble. But it was hard to make Bill Royal stick to this rule. Dr. Marwin and I found numerous huge fossil clamshells about five inches across. The bottom was soft mud with a layer of slimy travertine over it; every time I reached into it to pull something out, I stirred up the mud. I uncovered what appeared to be a smooth round white ball lying in the mud and pulled it out. It was the *acetabulum* of a human femur. I also found a sizable vertebra that belonged to an alligator. Dr. Marwin and I swam down to the vertical drop-off, but it became so dark we could see nothing. I looked out toward the center of the hole and then spotted the stream of large silver bubbles expanding as they came up from the darkness where Bill Royal was. I swam out and started to follow the bubbles down into the blackness, but the sensation became too eerie.

The second time, I went down with Bill Stephens. I was more familiar now with the edge of the slope. Bill and I went deeper into the hole, following the vertical wall down about 10 feet, and there found a ledge hanging out. Beneath it was a cave. Bill went under the ledge into the cave. His bubbles weren't trapped in the roof of the cave, as is usual, but came out above the porous ledge in numerous

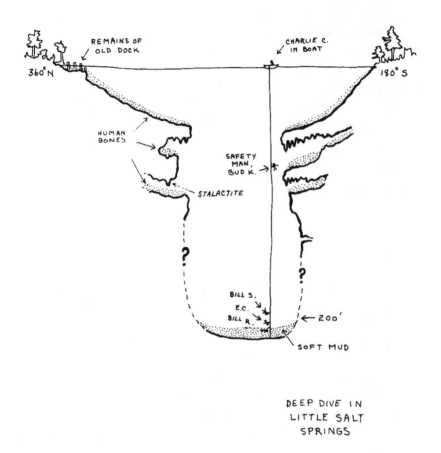

DEEP DIVE IN
LITTLE SALT
SPRINGS

thin streams. I followed Bill into the cave. Inside I could make out only his face and the silver bubbles from his regulator. I reached along the ceiling and felt stalactites hanging from it. The walls of the cave were slimy. As my eyes grew accustomed to the darkness, I could finally see the dim outlines of the entrance of the cave and I worked my way back to it.

I had just enough weights on my belt so that I could swim effortlessly up or down. I crawled out of the cave and slowly up the vertical wall. I looked up but couldn't see any daylight. I looked below me and saw a faint light. At first I thought it was Bill Royal deep in the hole with his light. Then I noticed a stream of bubbles coming from the surface and heading down toward the faint light. Suddenly I realized "down" was up and I was looking at bubbles rising toward the surface. When I had thought I was crawling up the vertical wall, I was actually heading down into the central sinkhole.

When I was properly oriented, I saw Bill Royal's yellow tank of air below the stream of bubbles. I hung suspended near the vertical wall and watched Bill coming up from the deep hole. He didn't notice me in the dark. He went right by me, drifting slowly upward with his lamp in one hand close to his face. In his other hand, he was holding some bones and examining them in the lamplight. The whole vision struck me as unreal and dreamlike. I was reminded of Alice falling down the rabbit hole and grasping a jar from one of the shelves on the side of the hole and reading "marmalade" on the label. As Bill continued upward, I became aware that I was now the person deepest down the hole. I started for the surface and soon saw the outline of Bill Stephens, who was searching for me. I joined him and we slowly swam to the surface, which seemed a long way off before we reached it.

Dr. Pettay didn't go down deep because her ears began to hurt at 20 feet. Bill Royal had been exploring the deepest caves and ledges. He was bringing up more human bones to Dr. Pettay, who stayed on a floating rubber raft. They had a total of seven femurs, a piece of skull, and numerous other pieces of bones.

Dr. Pettay thought the bones came from the prehistoric Indians who built the Indian mounds so popular in Florida as digs for both amateur and professional archaeologists. Bill Royal was convinced that the bones he found in the caves belonged to prehistoric people of a much greater age—thousands of years older than any other remains known in Florida.

Bill Royal, Bill Stephens, and I spent hours discussing Bill Royal's fascinating theory about an unknown ancient people, the first human inhabitants of Florida who lived in the caves when the water level in Florida was much lower.

Dr. Pettay and the other scientists who were familiar with archaeological and anthropological finds in Florida were more conservative in their speculations. Alligators could have put the human bones back in the caves. But the two Bills and I were convinced that alligators didn't dive to 70 feet.

The next time we dived in Little Salt Springs was March 18. It rained almost continuously the whole day. From the time I got up that morning, I expected someone to call and say the diving trip was canceled. But we had all made such elaborate plans for the trip no one felt like backing out. It's not like planning a diving trip in the Gulf, where weather conditions could make it impossible. You can always dive in a spring.

The difficulty in reaching the springs was magnified because of the rain. The road was half underwater. Bill Stephens, Bill Royal, and I transferred our gear into two jeeps the other diver had brought. I was wet and cold before I started diving, and I looked forward to getting some of the mud off my clothes.

The men made fun of me as I prepared to dive in blue jeans and a long-sleeved T-shirt: "Weeki Wachee Springs has its glamorous girl divers, and look what we've got!" Then Dr. Pettay took off her long raincoat and rain hat and brightened the day in a pretty Hawaiian two-piece bathing suit—red with large flowers. But I didn't have the only silly-looking diving outfit. Dr. Saunders, a veterinarian from Crystal River, was wearing some bulky clothing under a light tan union suit. When he wasn't diving, he had a feminine-looking robe that he wore over the almost indecent union suit. The robe was gray with a small maroon print and lined with imitation gray mink, which turned out around the neck to form a collar.

While we stood in the rain making plans for our dives, Bill Stephens took pictures of us huddled under Charlie Corneal's umbrella. It was finally decided that the first dive would be the one to the bottom while the air tanks were full. We were anxious to find out if the bottom of the hole was full of human bones, which would support one romantic possibility that the Indians used this spring to bury their dead or maybe even as a sacrificial well. It was decided that three of us would go all the way and that a fourth diver would go partway down as a safety man and stay at a depth of 100 feet with a spare tank of air in case anyone got into trouble. We took a small fiberglass boat out to the deep part of the spring and dropped a line down to the bottom.

Charlie Corneal stayed in the boat with more spare tanks. Bud Kraft was safety man. Bill Royal went down the rope first. I followed directly behind him, and Bill Stephens followed me. Bill Royal wore only bathing trunks and a white T-shirt, his face mask, flippers, weight belt, a depth gauge on each wrist, and a double tank on his back. I wore a shirt, tights under blue jeans, and woolen socks and sneakers under my flippers to keep my feet warm. I wore four pounds of weight, had on my white face mask, my *Aqua-Master* regulator, and a borrowed double tank. Bill Stephens wore a foam-rubber wet suit, and had a face mask with corrective lenses. He had a single tank of air on his back and carried a spare tank with regulator under one arm.

We started down. When we reached 50 feet, it was black all around us. Perhaps because my eyes were not adjusted to the darkness or because it was so overcast that day, I could not see the ledge at 60 feet when we passed it, but I felt it because our rope was right against it. Bill Royal was carrying a strong light and Bill Stephens another. When we got down past 100 feet, Bill Royal's light went out and he started coming up the rope. He passed us and went back to the surface. Bill Stephens and I started heading back up the rope, too, but then we met Bill Royal coming down carrying a smaller flashlight. We read his depth gauge. We were at 80 feet. Bill Stephens handed me his strong light and I shone it on Bill Royal as we started down again. My ears cleared easily and I hardly felt that I was making a particularly deep dive. The three of us stayed close together. I pointed the strong light ahead of me and on Bill Royal, and every once in a while I would flash it up to make sure Bill Stephens was following behind. Then the three of us stopped and checked Bill Royal's depth gauge again and read 130 feet. We dived deeper, and then the two Bills came beside me and we paused to look at the gauge on someone's wrist again. I couldn't make out what time it read, wondered what time it was supposed to be, and then decided I didn't care. I shone the light on my face and then on each of their faces. We were all smiling broadly.

As we continued down the rope, Bill Royal leading with his little flashlight and me lighting him with the strong light, I began to feel that I was breathing exceptionally fresh air. When we were deeper still, I suspected someone had opened the window in our stuffy dark room and fresh air was coming in. The air seemed to be loaded with oxygen and I felt very light. The air seemed to be flowing in my mouth without my having to draw for it. This very fresh air was rushing into my lungs so fast that it was an effort to stop it and exhale. At this point, the water in front of my light became extremely murky. Mud particles were dense in the water and I could hardly see Bill Royal in front of me. I kept the light on his hand that was sliding down the rope. The fresh air was now gushing into me, and there was a slow pulsating rhythm to the way it was flowing and the way I was thinking. It reminded me of someone saying a familiar phrase over and over to me.

I was sinking very slowly now into what felt like velvet ooze. I shone the light below me. Bill Royal had disappeared. Then I saw a hand sticking out of the mud. It was holding the rope. Millions of black particles danced in the short beam of light. The hand didn't move. I barely made out a gauge on its wrist. Oh, it's Bill Royal's hand, I concluded after a great effort of concentration. And Bill is dead, I decided. It was not a disturbing thought. I felt a rather calm sweet sadness. I attempted to reach out and touch his hand as I'd once, in a huge church, seen the man in line in front of me make a parting contact with a corpse in an open coffin. But I could not feel the contact. Suddenly the rhythmical sensation became intense and overpowering, and my head was going around and around. I imagined pleasantly, I'm having a baby and I want to go to sleep. The rhythm became a voice I'd heard before that said, "Take a deep breath and we'll all go together now, take a deep breath and we'll all go together now." Actually the voice had said only, "Take a deep breath." It was the experience I had in the delivery room at Buffalo General Hospital when my first child was being born and I was losing consciousness from the anesthetic coming in the face mask. Suddenly I realized something was quite wrong with my thinking. I couldn't be having a baby. I was diving and becoming anesthetized! I tried to clear my mind. I looked upward to see if Bill Stephens was above me. He wasn't there and the motion of turning my head made me extremely dizzy. The words "nitrogen narcosis" came into my mind, and I put my hand over my mouthpiece and held it tightly against my mouth. I must have remembered something that has always been in my mind since reading about nitrogen narcosis—that a diver is apt to take the mouthpiece out of his mouth and think he can breathe without it. My other hand was still holding the rope. "I mustn't let go," I told myself. I slid along the rope as fast as my flippers would lift me. The strange rhythmical sensation was beating in my head, and I thought I would lose consciousness any moment. I think I closed my eyes, because I don't have any visual memory for the next few moments; in fact I don't remember carrying the light, but it must have been in my right hand with the handle against the rope as I continued to climb it.

Soon I could think clearly again. The strange rhythm had stopped and so had the sensation that I was taking enormously long breaths. I had no idea how far up I'd come. I looked all around me with the light I was still holding. Bill Stephens had either gone toward the surface or had left the rope. I hoped it was not the latter, for we had all agreed that at no time would we ever let go of the rope. The rope meant so much to me—the only tangible object I could see in vacant blackness. I ran the beam of light along the rope as far as I could see in both directions, and then I paused and had a sickening feeling of confusion as I looked at the rope I was holding between my hands in an apparently horizontal position. I'd lost track of which side went up. I felt terribly alone in this dark no-direction nowhere. Then I remembered Bill Royal advising me, if I ever got into such a predicament, "Take off your weight belt, hold it out in front of you, and if it hangs upward, swim down to reach the surface." I wrapped my legs around the rope tightly and was about to take off my weight belt when I decided to feel for the bubbles coming out of my regulator instead. I reached in back of my head. The bubbles were going off to my left so I crawled hand over hand in this seemingly sidewise direction, which only logic told me was the way up. Then I thought of Bill Royal below me and how still his hand was on the rope. I felt terrible that I was not able to help him or stay with him. I decided to go back down the rope for Bill Royal. I crept down hand over hand what seemed a long distance. I was frightened to be alone. Then I started to feel again the sensation of taking very long breaths of exceptionally fresh air and was afraid I would succumb to nitrogen narcosis again. I wanted to get out of this awful depth and started up the line as fast as I could go, feeling cowardly and thinking how terrible it must be for Bill Royal down there by himself, sunk in mud. It did not occur to me to decompress. I only wanted to get back to where I could see something other than the rope. It seemed such a long swim before the darkness was broken and I saw a figure hanging on to the rope. I assumed it was Bill Stephens, but when I looked into his face I saw the man was wearing an ordinary face mask without corrective lenses. I was surprised and wanted to ask. "Who are you?" Then I remembered it must be the safety man, Bud Kraft. I had met Bud only that morning. It is sometimes difficult to recognize even a familiar face the first time you look at it in a face mask underwater. There were many large drops of water on Bud's forehead inside his face mask and he looked strange, but I have never before been so comforted by the sight of a face. In a similar situation on land, I might have grabbed and kissed him, I was so grateful to see another person. But such displays are hardly possible when you are wearing face masks and breathing through scuba gear. I motioned to him that I was going up, and then I pointed down, hoping he would realize Bill Royal was still there. Then I swam for the surface. The water became tinted with a green- light area above me.

As I approached from below the ledge next to our rope, I saw long stalactites outlined against the dim green light above—a sight which rather surprised me, as I knew this ledge was at least 60 feet below the surface, and on other dives I couldn't see such clear outlines at this depth even when looking toward the surface. My eyes must have been markedly more adjusted to darkness from this deep dive. I passed the ledge quickly without stopping to examine it.

I was so glad when I could see the bottom of the boat. When I surfaced, I saw Bill Stephens in the water near the boat. Charlie was standing in the boat. "I got nitrogen narcosis!" I said to Bill, and he said, "So did I!" I saw frothy blood on his face and in his face mask. He was bleeding from his nose. I told them Bill Royal was in trouble at the bottom. As Charlie was about to go after him, Bill Royal came to the surface with Bud Kraft.

There had been no need to worry about Bill Royal. I should have remembered he had been down to depths over 200 feet many times and had never suffered from nitrogen narcosis. He had felt no effect of it on this dive either. He told us that when he reached the bottom he held on to the rope with one hand (the hand I was staring at) and let the rest of his body sink into the ooze and mud. With his free hand he was groping around in the mud trying to find bones.

The person who had got into the worst trouble was Bill Stephens. At nearly 200 feet, he knew he had nitrogen narcosis. At the same moment, his single tank ran out of air.. He tried to change over to his spare tank, but it seemed an insurmountable task for him to change mouthpieces. With great effort, he finally did but then he lost his grip on the spare tank, and he swam for the surface bolding on to the hose with the tank dangling below. Bubbles of air seemed to be gushing all about him and he thought it was the effect of nitrogen narcosis. When he passed the safety man, it didn't occur to Bill to stop and check with him. Bill said he felt panicky because everything seemed so wrong. And we now saw why. The accordion-pleated hose on one side of his regulator was torn open at a place where he repaired it the night before with some liquid neoprene.

We rowed the fiberglass boat over to shore and warmed ourselves next to a fire the others had built. Bill Royal wasn't even cold, but Bill Stephens and I were shaking, either from the cold or fright, or both. Mrs. Royal had a thermos of hot tea and gave us all a drink. Our dive had taken twenty-five minutes. We never really got warmed up. The air temperature was colder than the water temperature, so we went back to our diving as soon as we were rested.

We made several more dives down the slope and recovered many more bones. Some were obviously recent bones—lightcolored and light in weight. The other bones were a dark reddish brown, heavier than natural bone and obviously fossilized. Many were human. Bud found a complete lower jaw with teeth. The men compared it to their own faces and decided that the man to whom this jawbone belonged must have had an exceptionally large face.

Bill Royal made the most spectacular find of the day. Back about 20 feet deep in the cave and some 100 feet below the surface, he found a human skeleton with a huge rock on top of a leg bone, as if the person had been pinned under the rock. Bill didn't disturb any of these bones; he wanted to get photographs before anything was disturbed. He offered to take Dr. Pettay, Dr. Marwin, and me into the cave to see it. The four of us went down the vertical wall to the entrance of the cave. I carried the light and kept shining it along the rocky wall and then at the other three divers, in turn, to keep a check on them. Bill Royal was leading the way and holding Dr. Pettay, a less experienced diver, by the hand. Dr. Marwin and I swam close by. We had no rope on this trip. Long stalactites were at the entrance to the cave. When I shone my light on Dr. Pettay's face, I saw some water in her face mask. She had an intense look in her eyes. Then saw that she was pointing upward and tugging at Bill Royal, who started taking her to the surface. Dr. Marwin and I followed until it was light again and we could see without the flashlights. Dr. Pettay was breathing normally through her regulator. Bill Royal continued up to the surface with her and Dr. Marwin, and I dived back down again to the ledge. We learned later that Dr. Pettay had been very uncomfortable when we reached the cave and she started to lose her sense of which way was up and which down.

Bill Royal dived down again, and he, Dr. Marwin, and I continued into the cave. I not only lost my sense of up and down, but I could not even figure which way was in or out of the cave. I marveled at how Bill Royal kept his sense of direction. He knew exactly where he was going and he kept following a ledge toward the back of the cave. (He told me later it wasn't a ledge along a side wall, as I thought, but a groove in the floor of the cave.)

In spite of the murky water, it was possible to get a good look at the cave walls (or ceiling or floor—I couldn't tell which) with our flashlights. The irregular surface of the walls had a pale orange-brown color by flashlight. I was sure I couldn't find my way back without Bill and wondered how Dr. Marwin felt. I flashed my light at Dr. Marwin and saw he was trying to tell me something. He gave me the signal that told me his air was almost gone (the index finger drawn across the throat) and indicated he wanted to go back. I turned to signal Bill, but instead Bill grabbed my elbow, turned me around firmly, and led Dr. Marwin and me out of the cave. I was disappointed that we didn't reach the body pinned under the rock. Bill told us that we were within a few feet of it but that he had trouble with his regulator and that it plugged completely in the cave. That's why he turned me around. Bill Royal could get no air at all from his tank, yet he had calmly escorted us the long distance out of the cave and to the surface without breathing.

After this dive, I was too tired to make any more deep dives that day. We dived around the top of the ledge some more. Bud found a fossilized antler with the side branches cleanly carved off—the first evidence of human craftwork.

The rain and mud and cold discouraged our group (except Bill Royal) from further dives in Little Salt Springs. We switched our diving to Warm Mineral Springs. Bill Royal reasoned that if early man was living in deep caves in Little Salt Springs when the water level was lower and the caves dry enough for stalactites to form, these early people must also have lived in the caves at Warm Mineral Springs.

The first cave-like ledge down at 35 to 40 feet below the surface was easy for us to explore. The sedimentation rate must be much higher in Warm Mineral Springs, for the cave floors had sediment about six feet thick before we reached the rock bottom of the cave. We were thrilled to find sedimentary layers, clearly distinct, indicating they had formed over many years—probably thousands. There were three major layers.

The uppermost layer was about three feet thick and composed of extremely soft mud with *Amnicola* snail shells throughout, as well as some alligator bones. Fragile travertine deposits formed loose chunks and coated stalactite fragments.

The middle layer, about one foot deep, was dark gray and compact. It had no *Amnicola* shells in it, but the shells of two different snails, *Helisoma duryi* and *Physa cubensis*, were heavily concentrated in the upper portion of this layer. Again we called on Dr. Clench for identifications. In this layer, we also found closely packed leaves of land plants, many pieces of charcoal, bones of deer and birds, and, as Bill Royal predicted, human bones. Bill, who made many more dives than Dr. Pettay, Bill Stephens, or I, also recovered two long bone needles, an antler-shaft wrench or atlatl weight, a bone pestle, part of a fossil shark's tooth with a chipped edge, and numerous other artifacts made from deer bones and antlers.

A bottom layer, averaging about one and a half feet thick, was light tan in color and contained pine cones, wood and charcoal, bones of turtles, birds, rodents, opossum, raccoon, deer, and many human bones, including the remains of a child.

Clay covered the cave floor. It was of hard limestone, with fossil remains of starfish, clamshells, and shark's teeth. We found a partly burned log three feet long, one end of which was embedded in the hard clay floor. Human finger bones were a few inches from the log.

We felt the find was an important one but could not convince any Florida archaeologist or anthropologist that it was worth investigating. Dr. John Goggin—the noted "Father of Florida Archaeology" at the Florida State Museum and the only archaeologist in Florida who was a scuba diver—disappointed us by being too busy to examine the site and did not think we knew what we were talking about. "Better stick to fishes," I was told by some of my scientific colleagues.

It was a frustrating experience. By now I was convinced that Bill

Royal had made an important find.

I told the whole story to Dr. Carl L. Hubbs, a noted ichthyologist who had also done work in paleontology and archaeology. He persuaded Dr. Suess at the radiocarbon- dating laboratory at Scripps Institute of Oceanography to run a carbon dating on a piece of the burned log. It was 10,000 years old! The human bones in the sediment with the log, if we could confirm that the layers were really undisturbed sedimentary layers as we firmly believed, would then be the oldest human remains known from the Western Hemisphere.

We thought that surely this carbon dating would bring qualified archaeologists to study the site, but it was difficult to find one who could even confirm the sedimentary layers. Finally, Dr. Harry Shapiro, the Curator of Anthropology at the American Museum of Natural History, who thought it might be worth investigating, encouraged a scuba-diving colleague of his, Dr. Ford, to check the sedimentary layers. Dr. Ford stopped off to see Dr. Goggin on his way down. Dr. Goggin felt certain that the human bones just drifted down through the sediment and were not in undisturbed layers of chronological age. He was anxious, however, to hear if Dr. Ford could confirm our claim of clearly demarcated sedimentary layers. Dr. Ford had one day to dive with us, and that morning he developed a cold. He attempted the dive but could only clear his ears to 20 feet, while we waited at 35 feet to show him the evidence. Word quickly spread that Dr. Ford, a qualified anthropologist from New York, came to dive with us at the springs but could not verify the sedimentary layers. We went around desperately trying to explain about his cold, but the people we wanted to listen to us had lost interest.

But something happened to us that made things even more

frustrating than Dr. Ford's plugged ears. The Huntley- Brinkley news program heard about the finds in the springs and wanted to report it on TV. Bill Royal was so pleased. With bright lights and good movies on a national TV program, everyone watching could see the sedimentary layers, and diving anthropologists and archaeologists would flock to study the important finds. But, as it turned out, Bill Royal was just too unbelievably lucky.

John Light, an expert TV photographer, was assigned by the Huntley-Brinkley show to do the movies because he had some scuba-diving experience. Gigantic underwater lights set the stage on a ledge at 35 feet. John, holding a huge expensive 35-mm movie camera, dropped down into Warm Mineral Springs to join Bill Royal and other divers waiting for him on the ledge. To their amazement, John missed the ledge and dropped like an anchor into the black depths. They had to go down after him, and they recovered him, still clinging to the overweight camera, at 170 feet. John confessed it was the first time he'd "dived" below 30 feet.

The rest of the dives went flawlessly—too well. John wanted Bill Royal to point out some recognizably human bones in the sediment. Bill cleared away a new place and found an entire skull, and the skeleton with it, lying in the middle of the three clear sedimentary layers. Even if this layer were a few thousand years younger than the bottom layer, it would be thousands of years old. As John Light took the movies, Bill removed the skull from its site. He posed with the skull next to his face, with as broad a smile as one can make underwater without leaking water into one's face mask. He turned the skull all ways, then upside down and noticed something white in the big hole at the base of the skull. He stuck his finger inside and found the white stuff was soft. They surfaced and took more movies

179

of Bill shaking out pieces of the "brain." Then they put the skull in a bucket of spring water and drove to the Lab, where Dr. Penner and I were dissecting some triggerfish.

Bill was so excited he could hardly talk. "We got a brain—thousands of years old a human brain—got it on the TV film—from the middle layer—thousands of years old—a fresh brain—look at it—feel it!"

We got a strong light, and Dr. Penner and I took turns looking into the *foramen magnum* while holding the skull near the surface of the water in the bucket.

"My God," Dr. Penner gasped. "It looks like a brain as fresh as you'd find in a butcher shop!"

As we gently tipped the skull, the white mass rolled around and we could plainly see the convolutions of the two cerebral hemispheres and the smaller convolutions on the cerebellum. We just couldn't believe it.

I phoned Ilias and some other doctors in Sarasota to get their opinions. They all thought it impossible for soft brain tissue to be preserved for any length of time. After a few hours, the brain started to break into pieces. I added formalin to the water to harden it, and it turned from white to gray. The next day, we took the skull to Ilias's office and he sawed a big opening in the back of the skull, and we put the brain fragments in a jar of formalin. Dr. Benjamin H. Sullivan, a neurosurgeon in Sarasota, confirmed our initial observations that some pieces appeared to be cerebrum and one large piece the cerebellum. From a gross examination of a block cut through a cerebral fragment, Dr. Sullivan could differentiate the outer gray matter of fairly homogeneous structure from the white matter which was fibrous in appearance. Dr. John

Bracken, pathologist, made microscopic examinations of stained histological preparation and could find no cellular structure remaining preserved, but Dr. Isadore Chamelin, biochemist, found 17.1% cholesterol in the white matter.

We wrote up a scientific report and sent it to *Science* for publication. It was turned down—mainly on the basis that the report sounded too sensational, that brain tissue could not be preserved naturally for thousands of years, and that unless the bone itself was dated we could not even conclude the skull was of any great age as no archeologist had checked the site and confirmed the sedimentary layers.

In the weeks that followed, few believed the story about the brain in the ancient skull. We were laughed at and ridiculed by many scientists and doctors, especially when the Huntley-Brinkley program was televised. "Genie, you don't expect us to believe a brain thousands of years old was found in front of a TV camera!" Even Dr. Hubbs, who was so interested in our earlier finds, was sure the brain story was faked to make a spectacular TV story. After the Huntley- Brinkley program, many of the scientists who had heard about the finds in the springs and thought we might truly have the oldest human remains in the hemisphere were convinced the whole story was a hoax. Soft brain tissue couldn't be preserved. Only those who saw the tissue and worked with us on it knew it was not a hoax.

For over six months, we tried to find some evidence to back up our findings, and then we heard from Dr. Tilly Edinger, fossil brain expert at Harvard University, who wrote me that it sounded quite possible to her, even though she had never heard of a brain preserved in water. Many human brains have been found naturally preserved in Peruvian and Egyptian mummies,* and in cadavers recovered from peat bogs. Once when a group of cadavers had to be exhumed from a cemetery in France, the brain tissue was preserved, even though the rest of the soft tissues had disintegrated. Dr. Edinger sent me a long list of fascinating, obscure scientific reports on these cases.

Then Dr. K. P. Oakley, the famous anthropologist at the British Museum who had exposed the Piltdown Man hoax, wrote me that he had just found a brain of Roman age preserved in adipocere. I sent him photographs of the brain fragments, all our information, and a piece of the brain itself. He submitted the latter for an ash weight, which was 42.3%, suggesting some mineral impregnation of the brain tissue. Dr. Oakley, in a letter congratulating us on the find, stated that he firmly believed the material to be the remains of brain tissue. A copy of his letter, statements brought to light by Dr, Edinger, and a confirmation by Dr. Hubbs, who came to the Lab and examined the brain tissue—and our report was finally accepted for publication by *American Antiquity.*

The age of the brain was still questionable. Since the burned log was found in the bottom layer and the brain from the middle layer, the human remains from the middle layer were undoubtedly less than 10,000 years old. We were fairly sure that the human bones in the bottom layer were close to 10,000 years old, but we knew that indirect evidence by nonqualified people like us would carry little weight.

Bill Royal removed all the bones he found with the skull. A

---

* Brains have been found in mummies that did not include the full mummification treatment, which included extracting the brain by suction tubes up the nostrils. Apparently, relatives sometimes could not afford the full treatment.

The skull found in Warm Mineral Springs with brain intact was later dated to be between 7140-7580 years old.

method for radiocarbon dating of bone was just being developed, and we sent fifty-two bones that belonged with the skull to Monaco, where, through the help of Captain Jacques-Yves Cousteau, tests were made on them by Dr. J. Thommeret, Chief of the Radiocarbon Laboratory of the Centre Scientifique de Monaco. We had to wait several years for the final dating, which was done on the organic fraction of the left ulna and right femur. We were pleased to learn the brain belonged to a skeleton between 7,140 and 7,580 years old. Although there is evidence that man has inhabited the Western Hemisphere for over 12,000 years, the carbon dating reported by Dr. Thommeret is the oldest reported for any human remains in this

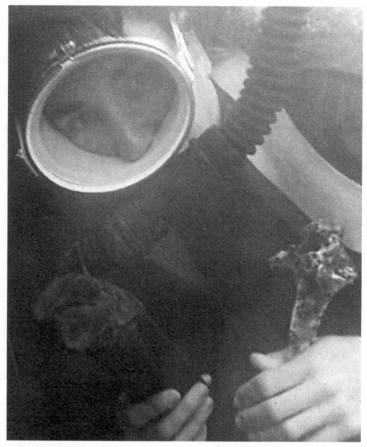

Coming up from a dive in Little Salt Springs holding a human femur found on a ledge.

hemisphere. And as this book goes to press, the valuable underwater sites at Warm Mineral and Little Salt Springs have still not been investigated by a qualified anthropologist or archaeologist.

11 More Educated Sharks

ALTHOUGH THE FASCINATING extracurricular activity of weekend skin diving in Florida's springs was scientifically a frustrating experience, concurrently I continued to have the satisfaction of working with fishes and having my ichthyological findings at the Lab readily accepted. Collecting and conducting experiments on sharks continued to be an important part of my work at the Lab.

We tried to train other species of sharks. Sandbar sharks, which we catch only during winter months, were easily trained after they got adjusted to captivity, but they did not survive when the water in our pens warmed up in late spring. Large tiger sharks are difficult to keep in healthy condition more than a few months. The adolescents do better. We managed to train a young seven-foot tiger shark to press a target marked with horizontal black stripes, in order to get food.

The big female lemon shark survived her mate (whom we apparently shocked to death with the yellow target) and, while pregnant with six of his offspring, went on with her learning. We also acquired another female lemon shark and called her simply L-3. This young shark, only six feet long and probably a three-year-old, had been run over by a boat. The propeller had cut three huge deep wounds across her back. We found her bleeding and weak in the bay near the Lab.

Oley placed her in the shallow end of the shark pen, where mud that settled in deeper parts couldn't clog her gills, and gave our old-timer L-1 a push if she came too close to the new sick shark. She recovered. I doubt if an adult shark could survive such an injury and maybe L-3 wouldn't have either, had it not been for Oley's tender care.

Our first experiments with L-1 and her mate L-2 had proved that sharks, at least lemon sharks, could be taught to press an underwater target in order to get their food. Now we could elaborate the experiment and test the ability of a shark to see the difference between different designs.

When L-3 was healed and seemed well adjusted to living in the shark pen with L-1, Oley and I set up the shark platform so we could lower two targets into the water at the same time. We also had two sandbar sharks in the pen, and started training all four sharks to take food from a white square target in the presence of a red circular target with the same surface area. As in our first experiment, when a shark took the food from the white square, his snout pushed the target, which swung back gently and closed the circuit, ringing a bell. If a shark happened to push the red circular target (held rigidly in place by a peg), the target did not yield, so the shark made a harder bump with the end of its sensitive snout; no bell rang, and of course there was no food reward. We switched positions of the targets from time to time so that the white square was sometimes on the right, sometimes on the left, of the red circular target, and the sharks could not predict which target held the food by the position of the target. We predetermined the initial position of the targets and the switches to be made for each test by having a student assistant toss a coin, a series of times, each day before the test began. Heads, the correct target was on the right (as the shark approached the two targets), tails on the left. In later experiments of this type, we used a table of random sequences. We got the mechanical kinks out of the experiment—adjustment to the changing tides (the targets were hung just under the water's surface), varying pressures on the targets as the different sharks approached and hit them in their individual ways, rusting of the automatic bell, etc. Oley and I even managed to keep our personal prejudices almost completely out of the experiments, and not to favor the frisky, scared, young lemon shark, whose exuberant dashes at the targets occasionally scored an error when we were sure she knew better. (You can't help getting to like some sharks better than others when you work with them every day.) We stayed objective and tallied her errors even though we felt her score was practically perfect. She seemed to be getting cues from the more experienced L-1. The young lemon was conditioned to working the targets in just a few days after she was presented with them and joined L-1 in removing food from the white square almost immediately, in contrast to the slower but eventual responses we got from the sandbar sharks.

We were able to make various tests on the learning ability and visual discrimination of the sharks. Sometimes our results were fuzzy because of the progressive ill health of some of the sharks. But L-3 gave us our most significant results.

After a few weeks of training, she was tested for her ability to distinguish between the white square and the red circular targets with no food cues. Her score in 12 testing periods was only 4 errors (pushing the red target), compared with 65 times she pushed the white target.

From this we concluded that she could see the difference between the two targets, but we did not know if she told them apart by the color or shape or both clues. I had purposely used two factors to make the targets unalike in order to increase the chances of a successful experiment in our first attempt to test sharks with two targets simultaneously. Now we had to break it down into separate factors. We tested her on shape alone, using a white square versus a white circular target. In 15 testing periods, she pressed the white square 56 times but made 41 errors by also pressing the white disc. Not only her poor score, but her general reaction to the targets (we couldn't excuse any of her errors as accidental) forced us to conclude that she could hardly tell the difference between a square and a circle, if at all.

But she readily distinguished between a white square and a red square. In 5 test periods, she pressed the former 18 times and never touched the red square target. But we could not be sure she did this on the basis of true color vision.

The shark, with its keen crepuscular and nocturnal feeding

habits and vision, might easily make out the differences in targets of different colors on the basis of differences in shades of gray or brightness. To eliminate brightness absolutely as a cuing factor, and prove color vision in an underwater animal, is a technical feat involving a knowledge of optics and physics. Some of our preliminary testing—and results of studies on other sharks and other species— made me realize sharks could recognize the difference between many different-colored objects, and I suspected this was on the basis of true color vision. But how could we be sure? Dr. Breder, who had watched with interest our various experiments involving color vision in sharks, had a brilliant and comparatively simple suggestion. (I had naïvely thought we could match the brightness of different colors by taking black-and-white photographs with an underwater camera until both colors registered the same shade of gray.) He proposed that we construct two targets so that monochromatic filters of different colors could be placed on the face of each target and behind this a waterproof box with a strong light whose intensity could be controlled with a rheostat. Then when we trained a shark to differentiate between, say, a red-lit target versus a yellow-lit target of what we estimated was the same brightness level, we could continue to test the shark by holding the light intensity the same behind the yellow target and varying the light intensity behind the red target from very dim to very bright. If the shark continued to choose the red target in preference to the yellow, regardless of light intensity, we would then have proof of color vision.

It was a neat plan. Oley and I started figuring out the construction of the boxes. I won't go into all our hang-ups.

Some we solved simply by further consultation with various experts, including Dr. Breder, whose knowledge of electrical and mechanical engineering continues to astonish his less versatile ichthyological colleagues.

It was the lengthy correspondence and my phone calls with optic supply companies and their consultants while trying to order the colored filters that wore me down and finally made me shelve the project. I discovered one can't just order a red and a yellow monochromatic filter. If you really want to know what you're doing, you have to know what specific narrow range of wave lengths you want, and in testing an animal's response to color (that is, spectral wave-length differences) you should first know where the peak of sensitivity is on the color spectrum for that particular species you are studying and under what conditions (water turbidity, amount of sun, clouds, etc., can shift this peak) you will be testing the animal. Perhaps after months or maybe years of testing lemon-shark vision with a complete range of filters covering all the wave lengths on the color spectrum and under conditions of controlled outside illumination and water turbidity, I might be able to have an idea of where the sensitivity peak is for the lemon shark and be able to order two color filters that would be a fair test of the shark's ability to see true color. Even then, maybe I would not really know if a lemon shark has color vision as man sees and understands colors.

Man sees certain wave lengths better in dim light than bright; his peak of sensitivity—in fact, the whole range of his sensitivity across the spectral wave lengths—shifts according to the amount of light under which he is viewing color. In very dim light, we lose our ability to see color at all and we see in shades of gray. Test your color vision by moonlight if you've never tried it before. Maybe a shark has color vision in dim light or a sudden peak of sensitivity at some wave length in, let's say, the orange of 585 to 647 nm and no sensitivity outside this range. Can you then say a shark has color vision? Suppose you happen to test colors outside a particular shark species' sensitivity range; you might erroneously conclude he had no color vision at all. The pitfalls are many.

The Navy and a lot of divers I know would like to know how much truth there is to certain rumors about the "yumyum yellow" popularly used in some makes of diving and sea- rescue equipment. I gave up my bright yellow flippers after my one diving experience under the phosphate dock in Boca Grande Pass. I had gone to see the giant groupers that are known to be under the dock in about 30 feet of water. Dock fishermen have exciting times there catching groupers up to 400 pounds. But Boca Grande Pass is also noted for its large hammerhead sharks, and we had evidence of sizes over 14 feet. So I stayed in between the narrowly spaced pilings as I dived during the short interval of a slack tide—and before I heard from another diver about how beautifully the hammerhead banks its wide flat head as it swims between the pilings! Luckily, I didn't see a shark, but my first sight of such a large population of giant groupers as lived at the bottom of the pilings was awesome. Several stirred and thumped like bass drums; most just lay on the bottom watching me—but not like other fish. These were looking not at my legs but at my flippers, which seemed to emit a yellow glow in the dark shadow of the dock. One grouper left its resting place and started to follow me, its head close to my flippers, as I swam for the surface recalling the gruesome stories I'd read of large groupers swallowing people. I sprawled quickly, if not gracefully, into the boat before the grouper got a yellow flipper into its mouth, as it seemed bent on doing.

I suspect the brightness factor of some of the luminescent- type colors is more important than the hue in the case of a number of dangerous fishes, especially barracuda. If a shark does see the bright color (as color or otherwise), it may still depend (in addition to a lot of other factors) on the particular kind of shark and his past experiences whether he will be attracted or repelled by the colored object.

I decided to leave that course of investigation to others. Some time later, I was interested to learn that is exactly what a number of scientists were studying at the University of Miami's Marine Institute. With much more ingenious and sophisticated methods than we were prepared to try, Professor Warren Wisby and some of his graduate students—especially Sam Gruber—strapped young lemon sharks into harnesses in tanks in a dark room and flashed colored lights into their eyes. After a series of special conditioning experiments in which electric shock, rather than food reward, created the conditioned response (eye movements), they concluded these sharks had color vision. In Japan, Dr. Tamotsu Tamura, of the Fisheries Institute of Nagoya University, and his coworkers (whom I later had the opportunity of visiting in Japan) dissected out the fresh retinas of some sharks and measured the electrical responses (electroretinograms) to wave lengths in certain color ranges, and decided sharks have color vision. Dr. Gruber and others have made histological preparations of shark retinas and found cells they consider to be cones—the receptors for color vision.

Yet today, after years of research along this line, other shark experts, as well as some of the best physiologists and anatomists in the world, have examined the evidence and claim color vision in sharks is not yet proved conclusively.

So it is with not too much regret that I look back at the switch I made in the experiments with L-3 when I decided to give up color testing for another type of visual discrimination experiment.

I was curious about L-3's poor ability to tell a square from a circle, and after reading about some fascinating experiments going on in Europe testing the remarkable visual abilities of octopuses, I had a possible explanation. The octopus investigators analyzed the image falling on the retina in terms of the distribution of the receptor cells or rods (if we limit ourselves to black-and-white vision, and let's do that for now) in rows of cells being stimulated by the visual image. I applied this type of thinking to the square-circle problem I gave L-3.

If you consider the square as the visual image hitting the shark's eye and being registered on rows of retinal receptors, the top row would stimulate as many rods as the second row, middle row, and all other rows right down to the bottom row, because in a square all rows would be the same width. Let us then suppose that in a more primitive optic system than ours, the stimulation each horizontal row of rods receives is sent back to the brain in a simplified manner in which the brain learns only how many rods in a particular row were stimulated, not which ones. In such a system, the animal would be able to distinguish between an "L" and an upside-down "L" in which the number of rods stimulated in the top rows (or the bottom rows) would be quite different in the two figures; but it could not distinguish between a "T" and the mirror images of an upside down "L" as all the horizontal rows in each figure would give the same message to the brain. The octopus has this limited type of visual discrimination of shapes.

Now let us compare the square and the circle. The

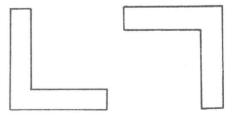

very top and bottom of these figures would register quite differently, but the numbers of rods stimulated per row in most of the two figures would not be very different, as the disc surface rapidly widens out

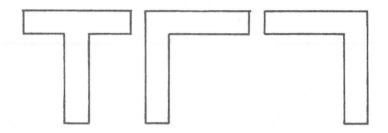

close to the width of the square.

I wanted to test L-3 on two targets that would have the same surface area (hence the same brightness), and would be similar in all respects except they would have a more dramatic difference in the numbers of rods stimulated in the horizontal rows. I chose a square versus a regular

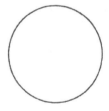

diamond, the latter being the same as the first except tilted at 45°.

The differences in each row between these two figures would be much greater than between a square and a circle, except at the very top and very bottom, where both diamond and circle would

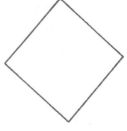

be practically a point and therefore equally different from the flat top and bottom square. In five testing periods, L-3 pressed the white square

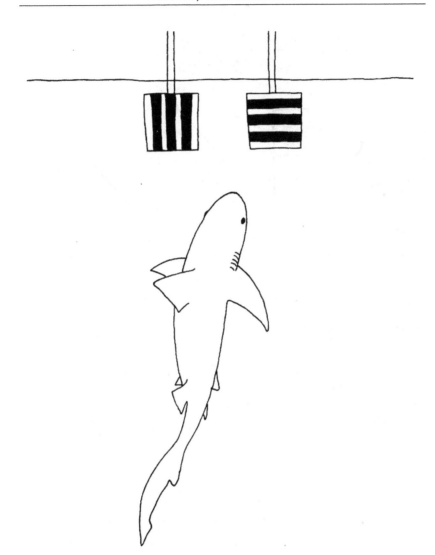

for her food 18 times and never once touched the white diamond.

I was curious about another design in connection with shark's vision. The little pilot fish, a member of the jack family of *carangid* fishes, is known to swim around the heads of sharks, even though there is no good evidence that they in any way pilot a shark. These pilot fishes have strong black vertical bands on their bodies, and

I figured any active swimming shark (as compared to the sluggish nurse shark that sits on the bottom and doesn't tour the sea with a bunch of pilot fish riding its bow wave) would be familiar with a design of vertical black stripes. Indeed L-3 had no difficulty distinguishing between the white square target and one that was like it except for having black vertical stripes an inch wide on it. The brightness factor wasn't accounted for in this case, however, and before I could test a vertically striped target versus a horizontally striped target of the same brightness, L-3 made her escape.

Experimenters who have worked with animals for any length of time will sympathize with the story of why only one shark, out of the four we started with, went through this set of experiments to test ability to make visual discriminations. Theoretically, we should have been able to carry out a whole series of visual discrimination tests on the four sharks we had. But the usual catastrophes, familiar to every animal psychologist, happened. In the long run, I think, people testing animals are more surprised if things go smoothly and all their animals live.

As the water temperature warmed up toward summer, the winter sandbar sharks died. During the summer, a mild "red tide"* broke out; it weakened both lemon sharks and finally killed L-1. Her six embryos, which had completed about half their development, were also dead when we opened her uterus. The young lemon shark survived. But a few months later we had the highest tide on record for the area; the water in the pen rose above the railing along the tops of the wooden pilings which formed the enclosure of the pen,

---

* A plankton bloom. In this part of the world, it is usually caused by a protozoan animal, a dinoflagellate called *Gymnodinium brevis*, which when it reaches high concentrations (over 100,000 organisms per quart of sea water) discolors the water and may kill millions of fish.

and L-3, beautifully trained by this time, swam out into the bay.

We wondered if she would enjoy her freedom after being accustomed to the regular and easy handouts she got from merely learning how to ring a bell. We also wondered if she might hang around the area, and for a while we watched for some sign of her and alerted the local stop-net fishermen about her. We didn't put an ad in the lost-and-found section of the newspaper asking for the return of a friendly young shark with three large scars on its back who answered to the ring of a house bell—but we thought about it.

12 New Quarters on Siesta Key

A BUSY YEAR interrupted our experiments testing sharks on striped targets. We had many other studies and experiments going on simultaneously, and even though we had added on several new rooms, our laboratory space was overcrowded as more and more visiting scientists from all over the world came to use our facilities to work on various marine plants and animals in our area, especially the sharks we brought in regularly. Our family had moved to Sarasota because my husband's growing orthopedic practice required that he live close to the Sarasota Memorial Hospital. We found a beautiful two-story Spanish-style house at the south end of Point of Rocks on Siesta Key. When I woke up in the morning, I could, without lifting my head from the pillow, watch porpoises swimming around the rocks in the Gulf. But it was an hour's drive each way for me to commute to the Laboratory in Placida. When the path of the Intercoastal Waterway was plotted next to our shark pens, we knew we would have to move the site of the Lab. The Vanderbilts suggested we consider moving the Lab closer to Sarasota to ease my commuting problem. On those days when a child was at home sick (and one out of four frequently was), I wanted to be closer to home. But real estate for a Lab site was extremely expensive in Sarasota. While we considered this problem, I had the most severe jolt and difficult experience I'd known.

My mother died from a brain hemorrhage in the summer of 1959. I lost much interest in my work, felt I could no longer handle a full-time job, and thought I should stay home with the children and help my stepfather, the person I knew suffered my mother's loss even more than I did. He had never mastered the English language, and with no other Japanese people around, as there had been in New York, he depended on my mother not just as a wife but as his only close verbal companion. It was going to be a lonely life for him running the restaurant without her, but he chose

Dr. Ilias Konstantinou who practiced medicine for many years in Sarasota and our son Tak

to remain in Florida with us rather than return to his Japanese friends in New York. I thought it would be better for him to move his restaurant to Siesta Key, within walking distance of our house. He agreed and we busied ourselves redecorating a rented store in Oakes' Plaza on Siesta Key into a Japanese restaurant. Even Ilias, who was busy with his medical practice, helped with the painting.

Nobusan or Pop opened the first Japanese restaurant in Sarasota at Chidori

Having never known my true father, Charles Clark, who died when I was an infant, Nobusan was my only father. However, at this time, I realized how much our relationship had depended on my mother's translations of the more subtle and complex exchanges between my stepfather and me. For the first time, we were forced to share a closeness in thoughts and emotions without our interpreter. It brought us closer together as we developed a way of discussing all things directly.

I wrote a letter of resignation to the Vanderbilts, telling them I couldn't devote the necessary time to the Lab any more but would stay until a new director could take over. Bill Vanderbilt got together with Ilias and Dr. Breder (now an official advisor to the Lab), and they proposed to me that we close the Lab temporarily until my family matters were more settled, and then perhaps I could go back to work on a part-time basis. I didn't think any laboratory should have a part-time director, especially a lab as busy as ours, but I agreed to try it.

Fortunately, a fine housekeeper, Geraldine Hinton, had just started working for us. When my mother died, Geri took on her responsibilities with a passion that reflected her warmth and sympathy and realization of how much I suddenly depended on her. Niki was only a few months old, and Geri took to him as if he were her own. All the children soon adored Geri. She managed them more with love and understanding than she did with strict discipline.

Geri was a handsome, buxom, light-colored Negro. After I told her she looked like Queen Liliuokalani or Hilo Hattie, she worked up enough courage to take Tak, Aya, and Hera (with their tans, they were darker than Geri) to the movies in Sarasota, still segregated at that time. If anyone ever questioned her, she planned to say, "If you insult people like this in Florida, I'm going back to Hawaya!" She told me this with confidence and glee, pronouncing "Hawaii" as no Hawaiian ever did.

After some months of difficult adjustment, my family's life was in a routine again, and my stepfather's restaurant in Sarasota was running smoothly and keeping him well occupied. Geri, who lived with us, had become a member of the family, and I found more and more time for my work. Within half a year, I was back as full-time

Lab Director. The big job now was the Lab's move.

Our small wooden buildings, built on skids, could easily be moved. But about that time our research work had begun to receive support from the National Science Foundation, and there were indications that we could get financial help from NSF to construct a larger laboratory. We talked to a real estate agent, Elizabeth Lambie, in Sarasota. She had heard about the work we were doing, knew of our nonprofit status, and talked the managers of the Palmer Estates into leasing us 81/2 acres of the most choice real estate on the south end of Siesta Key—a secluded bay-to-Gulf property facing mangrove islands in the bay and a minute's boat ride to Midnight Pass. It was an ideal location for our work and only a five-minute drive from my house. The land was leased to us at a nominal fee of $100 a year as a gesture of the Palmer Estate's contribution to supporting our work. Elizabeth Lambie, who would take no fee for arranging it, beamed triumphantly as if she had closed the most successful financial deal in the real estate business. The National Science Foundation awarded us a grant to cover half the cost of the new buildings, docks, and shark pens; and the Vanderbilts, in addition to all their regular support, gave the rest of the necessary funds for the modern new buildings of the Cape Haze Marine Lab which moved to Siesta Key, Sarasota, in the winter of 1960. Within a short time, our enlarged laboratory space was filling up with more scientists working on projects using elaborate equipment and instrumentation and taking up more space.

Dr. John Heller, who had seen the Lab grow from the first year and had seen the nature of my job change from one that was 3/4 research and 1/4 administrative or paper work to one that was 3/4 administrative, knew it was not a happy trend for me.

"What you need is a first-rate administrative assistant who can take much of the paper work off your hands, someone who can study catalogues, order with care the specialized equipment a lab needs, interview salesmen, meet visiting scientists and firemen at the airport, arrange the details of their visits, make motel reservations, and so on. You need a business manager!"

All he said, and he said much more, sounded just fine. But there was one big hitch. As the Lab grew, it cost more and more to run it. The research grants we obtained paid for new buildings, fancy equipment (such as a large refrigerated centrifuge), shark-fishing equipment, the bait and food to catch and feed sharks—even a second huge freezer to store special samples and a large refrigerator to keep perishable chemicals. The Vanderbilts had more than doubled the contributions they originally planned to give the Lab each year. Yet we did not have enough money for our housekeeping expenses—the phone, electricity (even nonprofit labs have to pay the high commercial rates), general maintenance bills, and the salaries of our staff.

We now had a full-time secretary, and she often had to have a part-time assistant to help maintain our sizable library; we had so many rooms and such large grounds that we needed a full-time maintenance man and a part-time yardman. As Ilias's medical practice grew, I no longer needed a full salary, and as I found I couldn't balance the Lab's budget any other way, I started cutting my own salary. Even this leeway would soon be nonexistent. So I asked John Heller just where he proposed I get the money to pay for the salary of an expert business manager.

"Simple!" he answered, as he often did to my most complex questions. "You don't pay him a salary."

John then went on to tell me there should be a flock of fine men living in Sarasota who have retired too early, are bored stiff with retirement, and would be delighted to contribute their services to work for a nonprofit organization as interesting as ours.

"What you need is a retired naval officer who can also help with all that paper work in connection with your grants from the Office of Naval Research and the National Science Foundation, someone familiar with the detail of government paper work that is starting to bog you down."

I couldn't believe the response we got just by passing the word around bank boards of directors and the social clubs in Sarasota that we needed a volunteer business manager. I felt embarrassed interviewing the fine applicants, administrators who could and had managed organizations many times the size of ours.

Colonel Rodney L. Carmichael, Jr., was the man who came to my rescue and kept me from drowning in a sea of back correspondence and other papers. He also took over the handling of an application we made to the William and Mary Selby Foundation in Sarasota, which contributed funds to build and equip three more laboratory rooms. One of these was equipped especially for the work of Dr. V. Z. Pasternak and her assistants, who were working on the biochemistry and extractions of lipids from shark livers in the burgeoning studies directed by Dr. John Heller. John had discovered a remarkable substance in shark livers he named "restim" (short for RES stimulator), which stimulates the body's reaction to fight and resist diseases many time above its normal ability. This reaction works through the RES, or reticuloendothelial system, and stirred great interest because of restim's ability to cause an alleviation and even regression in some types of cancers in Heller's experiments on chickens and mice conducted at his New England Institute for

Dr. Breder with his wife and colleague, Priscilla Rasquin.

Medical Research.

At one time, Dr. Heller had such a large crew of investigators at the Laboratory—including chemists and assistants from the Dow Chemical Company, who were developing a more automatic method to go through the twenty-eight steps necessary to extract the pure restim from fresh shark liver— that in addition to all our laboratory rooms we had to set up another laboratory under a huge tent. We rented a moving van, which we parked next to the tent, to hold all the chemicals needed in the operation. Colonel Carmichael took care of a thousand details arranging this setup.

Fortunately, because of our good weather in Florida, the tent (once plumbing and electric lines were run out to it) made a fine temporary lab room. Most of the time, we didn't even need to close the side panels, which kept out the hot sun yet caught breezes from all sides. And Dr. Pasternak calmly told us, as she looked over the setup, it was much better really than the concrete walls of our other lab rooms for ventilation and in the event of an explosion from the quantities of ether used in the first stages of liver extraction.

One of the innovations of the Lab at Siesta Key was an air-conditioned room with large tanks in which we could study the behavior of fishes and continue experiments on visual discrimination using small sharks under more controlled conditions. We could regulate—at least to a much greater degree than in our outside pens—the temperature, light, and water conditions, and could, of course, prevent sharks from swimming away at exceptionally high tides.

In these indoor tanks we discovered, to our amazement, the keen vision of baby nurse sharks, in contrast to the seemingly blind adults we first worked with in outdoor pens. It was some high school students who demonstrated how well baby nurse sharks could be used in studies of vision and the mental capacity of sharks.

13 Children and Travels

BECAUSE OF MY STUDIES in the behavior of sharks and the data I was accumulating on other large sea animals, I was invited to attend a number of meetings and conferences sponsored by the Office of Naval Research. One of these was held near Philadelphia in 1963. It gave me a chance to visit the newly opened Aquarama, where Curator Don Wilkie invited me to bring my children to ride the porpoises.

Niki, our youngest, was now four years old. He had exceptional breath control and complete lack of fear. He took the longest ride, hanging on to the dorsal fin of Star, a remarkable porpoise that seemed to know exactly when to bring Niki to the surface for air.

Niki learned to dive off a high diving board and swim underwater to the side of a pool soon after he could walk. He would frighten people. When he noticed he had an audience, he would dive off the board and, instead of swimming directly to the side, would swim to the bottom and hold on to the drain until people started diving in after him. Then he would let go and laugh bubbles all the way up to the surface.

From Philadelphia, we went to visit our long-time friends, the Kurokawa family, in New York. I took the four children and their young friends, Aiko and Zen Kurokawa, to the new New York Aquarium at Coney Island. Scuba-diving Curator Carleton Ray let them feed the walrus and other sea animals and brought the beluga white whales over to the window where we could watch the whales take fish from Dr. Ray's mouth as he dived with them.

My days with the children were sometimes exasperating. On one such day, I stumbled on a marvelous cure for a headache. I had just come back from a long day of driving to what turned out to be a poor spot for diving with children. After rinsing off and drying four squirming, screaming kids, I felt a migraine headache coming on. I was tired and chilled and decided to soak in a hot tub. Diving equipment, water toys, wet towels, and bathing suits were lying all around the bathroom. The place looked so messy, the children were arguing and fighting about some minor matter, my head started throbbing, and I suddenly wished I had the peace and quiet I've known at the bottom of the sea. The tub was filled near the brim. I reached for the nearest snorkel, stuck it in my mouth, and sank my head to the bottom of the tub. In a few minutes, my headache cleared miraculously and I could face the noisy tribe again.

Once, off the rocks in front of our house in Sarasota, hundreds of the practically harmless large moon jellyfish, *Aurelia*, floated close to shore. The children and I caught some in a dip net. Some jellyfish harbored over fifty young bumper fish,* like newly minted quarters, in this little understood symbiotic relationship. Niki, only a toddler then, was swimming with me when I pointed out this beautiful sight. I thought he would be awed still, and didn't expect his sudden reaction. He paddled quickly to the jellyfish, which was twice the diameter of his head, and pulled its undersurface right onto his face. He screamed and cried and I pulled him to shore. His face was red and rapidly swelling. I rushed him to his father, who gave him an antihistamine medicine and an analgesic balm, which quickly relieved his pain, especially on the tender part of his badly swollen lips.

I cuddled him in my arms. "Bad jellyfish!" I said.

"No," Niki said. "Pretty jellyfish was nice to little fishes. Niki wanted to kiss jellyfish."†

In July, 1960, I took the four children and their lovable nursemaid, Geri Hinton, on Sid Anderson's Bahama Adventure cruise. Sid had asked me to teach a course in marine biology to the diving teachers, nurses, doctors, and students. Geri assured me that I wouldn't have to worry about taking care of the children while I was diving and teaching. The children learned to dive among beautiful coral reefs on that trip, and we all took turns caring for a very seasick Geri, who hardly ever came out of her room unless we pulled into a port and she could debark.

---

* *Chloroscrombrus chrysurus.*

† The nematocysts or stinging cells of jellyfish, once fired, take awhile before they can be recharged with microscopic coiled poison barbs. It is best to hit even the relatively harmless *Aurelia* first with the tough palm of your hand before attempting to kiss it.

Hera saw her first shark underwater on this trip. We were scuba diving together with Stan Waterman, who was taking movies of a large green moray eel. Hera started swimming after a gorgeous colored queen triggerfish. I saw a shark about five feet long swim by her. She looked up and I thought she saw it, but then she went back to following the triggerfish. When we surfaced, I told her about the shark. "Oh, I saw it, but it was so small compared to the ones we dissect at the Lab. But Mommy, did you ever see such a beautiful triggerfish!" I had to admit we had none that colorful on the central west coast of Florida.

On this Bahama Adventure trip, I had a unique experience. I can get so absorbed in catching small fish in glass jars that I temporarily forget about the rest of the world. I often stay down more than an hour playing this rewarding game and making a collection in my net bag. Other divers, especially my children, get bored with my staying in one small area for so long, matching my wits with little gobies. So I would tie a string to my weight belt and put a cork float at the end. Then the children could snorkel-dive and come pull the cork when they wanted to go back to the ship.

One day after we came back to the ship, which was anchored near a staghorn coral reef for the day, others were still diving around the ship and I still had lots of air left. I sent the children aboard to check Geri, who was suffering as usual in her bunk. Soon I was absorbed in collecting fishes again. No one else was scuba diving now, but many were snorkeling and making an occasional dive to the bottom, about 30 feet. I had been down for some time when suddenly I felt a light tap on my back. I figured it was someone from the surface and looked up. There were a number of divers watching me intently. Three had movie cameras and were taking movies of me simultaneously, but I wasn't doing anything in particular and thought they were wasting a lot of film.

Then I felt another tap, and I turned and looked over my shoulder. An enormous brown eye, more than twice the size of mine, looked into my face mask. It was so close I had to back away from it to see what creature it belonged to. I moved very slowly as the eye kept watching me, and a flipper tapped me again. It was a huge old loggerhead turtle. Its carapace was encrusted with barnacles and one flipper was half missing, perhaps bitten off by a shark long ago.

The turtle came close to me again, and I realized he was looking at my glass jar as if he were hypnotized by it. I swam away very slowly, almost crawling along the bottom, holding one hand out and using the glass jar as a lure. The turtle kept following it.

I remembered diving with a native in the Palau Islands who hopped a ride on a turtle whenever he could. I had never been this close to one in the sea before, and decided to try it, I let the jar fall into the sand, and while the turtle looked at it I swam gently over his back, then quickly grabbed the front of his big shell with my arms, flattened my body onto him, and hung on tightly.

He took off, but not whiplike, as did the large nurse shark I once tried to ride. I was surprised at how easy it was to hold on and steer him. I rode him to the surface, then back to the bottom, up again and held him as the children went wild with joy and even got Geri to the railing to watch me ride the turtle.

I rode him around the coral reefs, and in a few minutes could bank and turn him through narrow passages. Stan Waterman wanted to get movies, and I was delighted to have an excuse to use another tank of air in this rodeo. We found it was easy to change riders. While I got a new tank of air, I turned the turtle over to Sid.

After we'd posed for all the pictures Stan and others wanted of us riding the turtle, Stan asked me for one more shot. He had missed me actually latching on to the turtle and wanted to try letting the turtle go and then photograph my catching him.

We planned the scene and then dived to the bottom. When Sid let go of the turtle, I was to try to get it. As Stan got his camera ready, I saw another turtle just as large, lying in the sand. It was wearing a silver halo around its head. Swimming up to it, I could see the halo was composed of many young bumpers, the fish in the jack family who young Niki and I had seen traveling in groups under the bell of the moon jellyfish.

But I've never heard of baby bumpers around a turtle's head. I had no trouble catching a ride on this second turtle, but the bumpers were distracting as they circled both our heads in a fluttering swim. This turtle was even older and more sluggish than the first, and seemed to have all its flippers a bit chewed off.

I was sorry we were scheduled to leave this strange place in the sea, where crippled old turtles found refuge and set up a symbiosis with bumpers. On subsequent trips, the children dived with me at many other Caribbean islands: Nassau, Bimini, Aruba, Curaçao. At Cozumel, we all got our first good look at a large hammerhead shark swimming around us, and Hera and our young guide Doug Peterson (one of the volunteer student assistants at the Lab) dived into coral caves and brought up black coral. Along the Florida Keys, especially Islamorada, Key Largo, Key West, and during several trips to the Tortugas, the children have dived and helped me collect fish.

A special treat for me was introducing my children to the waters of the Middle East in 1964.

On the way to Eilat on the Red Sea, we stopped to bathe in the warm and too salty waters of the Dead Sea and the cold, fresh mossy-banked stream in shaded crevices at Ein Gedi. In Eilat, which we reached at nightfall, we stayed at the Coral Beach Hotel, noted for its informal atmosphere. In the morning, we got our scuba equipment at Willy's handy dive shop next door before taking our first dive to see the garden eels on a sloping underwater drop-off near the hotel.

Before leaving Jerusalem, I had asked Dr. Heinz Steinitz, Israel's distinguished marine biologist in charge of the University's fish collection, if there was anything in particular he wanted for the collection. He requested I bring him a specimen of the elusive garden eel, known to be living at Eilat but never collected there.

The divers hanging around Willy's dive shop all knew where the garden eels lived. "But you'll never catch one," they assured me.

The children snorkeled at the surface and watched me go down to the eels. I wore scuba gear and took a hand net and a plastic squeeze bottle with formalin inside. There were over a thousand eels in this "garden," which spread along a grassy slope of sand bottom from 18 to 35 feet underwater. The anterior parts of the bodies of the eels were held in a vertical position and protruded from their burrows a foot or two.

Their long smooth bodies undulated gently and irregularly as they picked plankton from the water and swayed their heads from side to side, their large brown eyes carefully watching the surroundings. The ends of their two- to three-foot bodies were kept inside their tube retreats in the sand. As I approached within ten feet of an eel, it would start to withdraw into the sand, and each section of the garden slowly sank and disappeared as I swam over and got a close view of only the holes. The holes were less than an inch in diameter and spaced roughly one or two feet from each other and sprinkled over the sea slope. I knew from past experience it is hopeless to try to dig in the sand for a burrowing fish. So I tried squirting formalin down some of the tunnels, then backed away and hoped the irritating chemical would drive out the eels. After several minutes of waiting, I gave up and started swimming off when I heard Hera at the surface splashing the water and yelling to get my attention. Then she pointed to the area where I had put the formalin. A big long eel was staggering out of its tunnel, its head drooping from side to side. I swam over and grabbed the eel and pulled its hind end out like a robin yanking up an earthworm. When I got to shore, a crowd had gathered to see the first garden eel collected at Eilat.

The following day, as I dived past the garden eels to where the sea bottom slopes more abruptly and is of bright white sand with scattered coral heads, I saw a strange little fish hovering at an oblique angle a few inches above the sand. It held this suspended position with undulating movements of its elongated body. When it stopped undulating, it started sinking, then erected a plume-like dorsal fin with filamentous rays, whipping the tips forward into a fan shape and stiffening its body into a horizontal position before sinking again into an oblique position. When I approached closer, it dived into solid sand and was gone with no hole to mark where it had entered.

It was a rare type of sand diver belonging to a family of fish never reported from the Red Sea. This type of fish was occasionally caught in dragnets in the Indian Ocean but never seen alive. I scrutinized the sand where it had dived in and nearby saw two shiny round beads among the sand grains. I put my hand net over what I hoped were the peeping eyes of the fish, plunged my free hand into the sand below, and the fish, *Trichonotus*, jumped into my net. Niki was

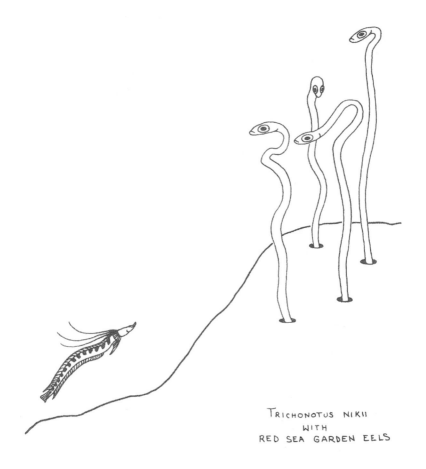

TRICHONOTUS NIKII
WITH
RED SEA GARDEN EELS

at the shore and we filled his face mask with water and put the fish in it until we located a bucket. It turned out to be a new species of fish with beautiful black internal "eyelashes" (a complicated umbraculum with hair-like attenuations radiating across the pupil from the iris), probably for cutting down the glare from the bright sand with which it so intimately lived. I named it *Trichonotus nikii*, after the youngest member of our expedition to the Red Sea.

219

Aya turned out to be the champion garden-eel catcher and eventually caught three of these eels at another garden farther south on our next trip to the Red Sea. Hera and Aya found night diving in the Red Sea a bit spooky but wouldn't turn down a chance to join me. Just after sunset, there was still enough light to dive among the corals as the nocturnal animals started to come out of hiding. Parrot fish wrap themselves in mucous blankets for the night, and other daytime-active fishes change their colorations and tuck themselves to bed among coral crevices or down tunnels in sand and sponges, all eyelidless but each sleeping in its own way. The crinoids, tentacled purple lilies of the Red Sea by day, crawl to the tops of the coral reefs and change from radial to bilateral symmetry, reshaping their feathered tentacles into spread fans to feed on plankton at night. And soon the bald pastel coral heads are wigged in moving tufts of hairs turning from purple to black in the fading light. By the time we have to turn on our underwater lights, *Gorgonocephalus*, the mysterious "walking bush" of Cousteau's movie—really a monster starfish—has crept out from hiding and stretched out the curled and wriggling ends of his hundred tentacles to catch his food; he is joined by fearless big-eyed orange squirrelfish, bronze sweepers that hid in caves all day, and moray eels—all on the prowl in the dark. Even stationary objects seem to move in our bobbing shafts of light.

I was so pleased that all my children liked to dive and dived so well, but I had to work at holding back anxieties. Even though Tak had won forty-two blue ribbons at swim meets that summer, when Bill Royal buddy-dived with Tak and both got out of my sight in the mud-stirred outflowing tidal water at New Pass, I was uncomfortable. When Bill took Hera down to 80 feet, I stayed on shore trying to look calm, but I could not take my eyes off the surface of the water until they came up. I hope someday I can watch my children make dives without me and feel at ease or unconcerned. Or is this attitude only a facade some parents learn to cultivate when reason tells them their children are old enough to be on their own?

14 An Imperial Ichthyologist

THE ARONSON FAMILY knew the Lab from the summer of 1956, when Lester first had brought his wife and two young sons Carl and Freddie to enjoy a Florida vacation in a beach house on Manasota Key while he worked at the Lab on the behavior of blennies and gobies. When Freddie went to high school, he entered some science fairs and walked off with several top awards and a scholarship for a project in which he did experimental behavioral studies on rats. He impressed his father with the diligent care he daily took of his large live rat stock while conducting the experiments. The following summer, in 1963, Dr. Lester Aronson brought sixteen-year-old Freddie to work at the Lab.

Freddie's patience in the care and training of young nurse sharks, during three consecutive summers, and his special talent and ingenuity for designing and making the apparatus used in our experiments formed the foundation for the studies we were able to complete and publish on the vision and learning ability of young nurse sharks. During these summers, Tim Wright, a high school student, first trained and conditioned baby lemon sharks at the Lab, then returned two more summers, as a student at Cornell University, to assist Freddie in the daily training of sharks. Doug Peterson, a high school volunteer from Texas, also worked with us and was the first to teach a baby nurse shark to make the discrimination between vertical and horizontal stripes. We made a fine movie sequence to add to our film on student experimentations at the Lab. Not only was I surprised by the ability of young nurse sharks to see well, compared with adults which had seemed practically blind, but I was surprised that the striped targets were as easy for the nurse shark to distinguish as they were for larger, more active sharks which have their entourage of striped pilot fishes. The more we tested sharks on various striped targets, the more we realized what a particularly easy pattern it was for most sharks to see.

Freddie Aronson's experiments with young nurse sharks culminated in a light-dark discrimination problem. The apparatus Freddie assembled seemed formidable to me when I saw the wiring involved—all worked out and built by this quiet boy. He built a plexiglass box with a shield in front and two openings near the bottom. He recessed small transparent targets behind each opening. Behind each target he placed a black vertical tube containing miniature lamps to illuminate the target directly in front but a black rear divider prevented light from reaching the alternate target.

The lights were controlled by an electromagnetic system which automatically shifted (or, on purpose, did not shift) the light from one target to the other in a programmed random sequence each time the correct target was pushed by the shark. Each push was recorded automatically on separate correct and error counters. In this experiment, the illuminated target was designated correct. The shark was rewarded with food if it pushed the correct target. Freddie trained his shark every day, and the Lab room was darkened during each test.

The little shark learned to make the discrimination in five days. The learning curve was practically identical to that of white mice tested on a similar problem.

During the last seventeen days of testing, the shark didn't make a single error. We liked to show him off to visitors and claim that this sluggish little two-foot shark was the brightest animal ever tested at the Lab. A more truthful statement is that this shark had the best teacher and was taught under the most carefully controlled conditions.

Shortly after the Lab moved to Sarasota, Oley left and joined Beryl working at the water plant of the Vanderbilt's real estate development at the Cape Haze Corporation in Placida. The long commuting to Sarasota was too hard on Oley, who was also changing over from the lonely life of a bachelor fisherman to that of a well-fed happy husband who looked forward to being home with his bride. Before he left, he broke in a new collector-boatman, John Strong. John looked like his name. He was almost a foot taller than Oley, broad-shouldered and husky compared with the wiry strength of his slim predecessors. Young children adored him and he enjoyed telling them about his unusual past experiences, such as being a hard-hat diver for an oil company in New Orleans, or showing them the latest trick he was practicing from the last meeting of the local magic club. His knowledge of boats and boatbuilding was impressive. When we needed a larger boat, he offered to design and build it for us, and I might have let him try except that his time became too valuable as our shark collector and the strong yet gentle arm in all our shark work. When I applied to the National Science Foundation for funds to build a 36-foot special boat for our shark work, I felt it was John who convinced the NSF representatives, when they visited the Lab, of the practicality and efficiency of the design we had chosen. The *Rhincodon*, named after the whale shark—the largest yet gentlest of sharks— was John's production. He got the boatbuilder, worked out the design and cost with him, and explained to me the advantages and disadvantages of everyone's ideas and suggestions for what we should and should not have on the new boat.

John could improvise most anything needed for a quick repair job in any emergency. He handled the *Rhincodon* admirably in rough weather. But he could also run it aground on a sandbar while daydreaming on a peaceful sea. And he could run out of gas several miles from shore in the Gulf on a routine check of the shark lines, or forget to let go of a big dead shark when he pushed it off the dock.

John usually worked barefooted, as many of us did at times in the informal atmosphere of the Lab. (Only once did we see "formal" attire, when the British endocrinologist Dr. Anthony Perks showed up to work with sharks in a long white lab coat, tie, socks, white Oxford shoes—and we don't have to go into what happened when his knife hit a large bull shark's artery.) Sometimes John wore "thongs," especially if we were expecting visitors, but often he'd end the day with only one on, as the other got mislaid, fell overboard, dropped into the shark pen, stuck in the mud, caught in a motor, or got its strap broken.

John was fond of wearing white T-shirts with wide black horizontal stripes, which gave his shoulders and chest an even broader look. He was also able to get newly captured sharks to eat readily. He spoke to me about how sharks "recognized" him, but I suspect a strong factor in this recognition was John's T-shirt. John trained a newly caught tiger shark to operate a striped target in the record-breaking time of three days. It amused us, in a Charles Addams sort of way, when the following week *Life* magazine came out with a cover story on the latest style in bathing suits—bold stripes! We didn't really think this would influence statistics on shark attacks of bathers, but if anyone asked my opinion about wearing apparel in "shark-infested" waters, I would certainly recommend avoiding the striped effect. Unless, like some divers I know, you would like to attract sharks.

Some of the strangest shark-protective wearing apparel for a diver that I've seen was presented to me, rather ceremoniously, by Professor Toki-o Yamamoto of Nagoya University when he came to visit the Lab. This beautiful old man looked like an Oriental guru—with a long thin beard, curly hair to his shoulders (which he sometimes wore in a net), and a most impressive thespian voice when he chanted to us a poem about Sarasota that he composed on the spot or sang "Santa Lucia" in Italian with his heavy and charming Japanese accent. Actually he is Japan's leading fish geneticist and came to the Lab to attend an international conference on inter-sexuality of fishes and to tell us about his findings on the genetics of sex determination in some Japanese fishes. He knew about my study of the little hermaphroditic *Serranus*, but he had also read the Japanese translation of my book telling of my love for diving in tropical waters. He decided I needed protection from sharks, and brought me an unusual diving garment, which he said was used by some of the Japanese ama divers when they thought sharks might be around.

This garment was supposed to be simultaneously both a repelling and an attracting device for sharks; the psychology of this is intriguing, for it takes into account sharks of all sizes, type, and intent. Essentially, it is a glorified G string. One part is a yellow, black, and red embroidered ribbon (less than an inch wide) that ties low around one's waist—or hips, if you want to go bikini. From this belt a brilliant red cotton cloth, six inches wide and eight feet long, hangs down in front to form a crotch strap, which can be looped through the back of the belt, and then trails behind the diver as a conspicuous tail. I had on my blue jeans and shirt when I tried it on at the Lab, and Dr. Yamamoto showed me how I should wear it. But I imagine it comprised the entire swimsuit when the famous Japanese women divers use this garment and secure their abalone knife in the belt.

Now, if a shark comes along and sees a diver in this outfit, there are three possibilities. He may be indifferent in his reaction, and therefore you have nothing to worry about. If he is a small shark or a large coward, this strange, eight-footlong, moving object will probably frighten him off. Unlike porpoises, with their highly developed method of communication, I don't believe a shark will go after a school of his companions and say, "Come with me to see this crazy fish," which could result in an in-union-there-is-strength situation.

In the event the shark is attracted to the diver, the role of the red G string is ingenious. The bright red cloth will be more conspicuous, flutter more irregularly, and therefore be more attractive than the body or limbs of the swimmer. The shark will then take its first investigatory bite on the trailing red cloth and, disappointed in the taste of the material, leave the diver alone.

Utamaro's woodcuts of Japanese ama divers collecting abalone go back to the 1700s; I have great respect for a shark-protective garment evolved from so many years of diving. I thanked Dr. Yamamoto and bowed back at him as I accepted his gift and caught the delighted twinkle in his eye. I've never had the opportunity to test the red cloth when a shark was around. In my imagination, however, I've executed a verónica as a shark charged the red cloth. I also have a vivid picture of myself diving under the phosphate docks at Boca Grande trailing the long red cloth, first feeling secure in the great distance between me and the end of the cloth going into the giant grouper's mouth; then, because of the vacuum-cleaner suction of a fish's large mouth, seeing myself being pulled into the grouper like an insect on the tip of a frog's tongue.

At the end of the summer of 1965, when most visiting scientists were returning to their own labs and universities, Freddie was getting ready to return to school and I was preparing to go to Japan. The busy summer activities at the Lab would come to a halt in September, and we normally had a time of comparative rest when we could get caught up with many items: writing reports, painting and repairing the buildings, docks, pens, boats, etc., during a month when the hurricane season often forced us to dry out our shark lines.

I had written ahead to a number of my Japanese scientist friends asking to visit their labs and giving them my proposed itinerary. Professor Yasuo Suyehiro, a most famous Japanese ichthyologist, wrote me that the Crown Prince, Akihito, of Japan would like to meet me and that I was invited to his palace in Tokyo. Professor Suyehiro was one of the Crown Prince's teachers. Prince Akihito, a young ichthyologist, had already published scientific papers on the anatomy of gobies and was following his father's interest in marine biology. The Emperor is famous for his collections of invertebrate animals and the beautifully illustrated books he has published on *ascidians* (sea squirts), *opisthobranchs* (sea slugs), and crabs, which he collected near his summer home at Sagami Bay. It is not surprising that one of his sons took up the study of fishes.

What does one do when one is invited to the palace of the Crown Prince? Knowledgeable friends told me, "You must take him a present."

I jokingly said to Freddie, after considerable serious thought gave me no solution for an appropriate present, "How about your trained shark? I can be sure he hasn't already got one." To my surprise, Freddie was delighted with the idea.

"I'll make you a portable testing apparatus so you can show the Crown Prince how the experiment works."

Freddie must have stayed up most of the night working on the portable apparatus. Dr. William Tavolga, who was winding up his summer's work at the Lab, helped Freddie with some of the finishing touches. They were intrigued with the idea of miniaturizing and simplifying the bulky apparatus so that it could be folded into a package I could carry on the plane. They didn't pay any attention to my protests and my audible worrying.

"But how am I going to take the shark on the plane with me all the way to Tokyo? Suppose it dies on the way? And what am I going to do, walk into the palace and say to the Crown Prince, 'Here's a trained shark for your salt-water pool'?"

I never got so much cooperation. Mrs. Pat Mathews at the Sarasota Travel Bureau was utterly delighted to handle my baggage problem. She'd never had such a challenging job to do. Everything was arranged, and in just a few days the shark and I were ready to fly to Tokyo.

National and Pan American Airlines sent special agents to help me at transfer points in Tampa, Los Angeles, and Honolulu, and gave me an extra seat at no charge for the shark, who got more attention from the hostesses than any passenger. Spencer Tinker, in charge of the Waikiki Aquarium, arranged for the shark to stretch out and have a swim in one of the tanks during our stopover in Honolulu.

I was greatly relieved, as we approached the airport in Tokyo, when I took a last peek into the carrying box (the size of a large hatbox), which was lined with several large plastic bags and filled with just enough water to cover the shark, and saw that it was still OK. I put the cover on, slung the shoulder strap over my head, and carried the shark off the plane in my giant shoulder bag.

My telegram to Dr. Suyehiro had created a sensation. I told him I was bringing a live small nurse shark for the Crown Prince, and would need a place to keep it until I could take it to the palace. I thought the news would surprise him a little but not cause the trouble it did.

A giant truck was waiting for the shark at the airport. On the back was a salt-water aquarium large enough to house a twelve-foot shark. In the midst of a huge welcome party, including friends, scientists, professors, newspaper reporters, TV cameras, and gawkers, we dumped the little shark that had traveled in a hatbox halfway around the world into the giant aquarium that was to transport him to the estate of Mr. Shinichi Saito, a friend of Professor Suyehiro's, who lived on the outskirts of Tokyo.

Mr. Saito's hobby is raising fishes, and in various parts of his estate, among exquisite Japanese gardens, he has buildings for housing aquariums for numerous fresh and saltwater fishes. Working with Japanese scientists, Mr. Saito is raising Russian sturgeon in a huge doughnut-shaped aquarium and finds he can grow them much faster than the Russians have reported. He also builds aquarium supplies, and his prize product was a machine, the size of a standard air conditioner, that could be hooked up to a large aquarium or series of small aquariums to control the temperature and condition the water. He planned to give one to the Crown Prince.

Mr. Saito had built a special aquarium and room in his garden just to keep the shark until it moved into the palace. His staff of assistants and servants helped me set up the testing apparatus, which Freddie had arranged so that batteries could he used for lights and an operator could control the lights manually. We tested the shark, who immediately pushed the target that lit up. He had had a long trip, cramped in the hatbox, but all the changes in aquariums and all the people who handled and petted and photographed him did not confuse him. His appetite was a little poor at first, but he pushed the targets correctly whenever one lit, even if he was not too interested in the reward food.

We learned that the Crown Prince had a room at the palace ready for the shark (we decided not to make the gift a surprise). We had the shark delivered to the palace in advance of our formal visit, so that it would have a little time to adjust to its new aquarium and palatial environment before our formal visit and the demonstration. I was escorted by Mr. Saito and ichthyology Professors Suyehiro and Abe. When we arrived, we were greeted by the Chamberlain, Mr. Yagi, who escorted us to a conference room where we were presented to the Prince. I had been asked to bring pictures of the fishes I had studied. The Prince, who spoke perfect English, asked many questions about my work and the colored slides I showed as I became less aware of his royalty and more aware of his intense interest in fishes.

Next we went to see the shark. It was in a low aquarium, 4 by 6 feet, set on top of a table in the middle of the room, so one could walk all around it and not have to bend over in order to get a good look at the shark. The room had been equipped with black draw curtains, and when the shark was ready to perform for the Prince, who knew in advance what to expect, one of the palace servants stood by holding a beautiful platter with slices of raw lobster arranged in the shape of a flower. When the shark pushed the target, he was rewarded with a slice of lobster placed before his mouth from the tips of long delicate chopsticks, inlaid with mother-of-pearl, held by the palace servant.

After the shark performed for the Prince, the Prince showed me his laboratory rooms at the palace containing aquariums in which were some of the gobies he was studying. Then, in a large American-style sitting room, we had tea and delicate cookies variously shaped and colored like fallen autumn leaves being blown and tumbled across the ground. While I admired the artistry and skill of the palace pastry cook, we continued our discussion of fish. The Prince confessed he didn't like to eat fish, only to study and collect them, and one of the main reasons he had chosen to study gobies was because he enjoyed wading along the shores of streams and beaches where he could catch them easily. I was surprised to learn that the Prince had never tried skin diving —not once looked underwater through a face mask. I assured him he was missing a wonderful experience. Our conversation ran to great lengths; as the designated duration of our visit ran considerably overtime, the Chamberlain rose during occasional lulls with such remarks as "The Crown Prince is indeed pleased you could visit him," but was cut off with a brief one-word comment in Japanese by the Prince—"Made!" (not yet)—who remained seated and then continued asking me questions. I didn't get up to leave when I knew it was time. I had learned, after an embarrassing situation in Ethiopia when my expedition colleagues were aghast, that one doesn't leave the room before royalty. So this time I made no motion to leave. I even told the Crown Prince of my friendship (in spite of my blooper at one of our first meetings) with Prince Alexander Desta, Commander of the Ethiopian Navy, who had immeasurably aided our expedition to Dahlak and who had been game to eat the raw giant clam, *Tridacna*, when we were diving in the Red Sea. The Crown Prince had also met Prince Desta in Ethiopia, and seemed to consider him lucky to have the time to go diving.

After my visit with the Crown Prince, I had the opportunity to visit many of the marine laboratories and aquariums in Japan and to meet the many hospitable ichthyologists and avid divers of Japan. My book *Lady with a Spear* had gone into eight editions in Japanese and a special English version with Japanese footnotes for schoolchildren studying English. In Oita especially, I was given the full treatment for a visiting scientist. I was assigned (after some deliberation, I suspect) a geisha girl during the evening I was the guest of honor at a dinner hosted by the Governor of Oita Prefecture. My companion sang, danced, and played the shamisen for me whenever I wished. She sat beside me throughout the elaborate formal dinner, pouring my tea and explaining the unusual seafood dishes the waitresses served. She didn't eat, but joined me in drinking sake, in which soaked the fin of a poisonous puffer fish. She would have smoked a cigarette, I'm sure, as the geishas assigned to other dignitaries were doing, if I smoked. She spoke English and lived up to the reputation of good geishas by making me feel at ease and comfortable with her pleasant conversation, which showed off also her special preparation. She discussed the women ama divers of Japan and their problems with sharks. She also knew I had done my doctoral dissertation on sexual behavior of fishes. Finally she asked quite seriously, "Is it really true that sharks have twice as much fun as we do?" I had to think a moment before I could explain that as far as we ichthyologists knew, it is physically impossible for the male shark to insert more than one clasper at a time during copulation.

Of course, when I went to Nagoya, I visited Professor Yamamoto—or Dr. Killifish, as he is called—and saw the enormous array of tanks where he breeds and raises killifish for his genetic studies. I was equally impressed by the number of students he was training in this field—each of whom obviously worshiped this master of genetics.

When I went diving with the Japanese women ama divers, which Dr. Suyehiro arranged for me to do, and when Dr. Abe took me to visit the Tokyo fish market (the largest in the world) and I saw, for sale, the rows of giant sharks, and the rare six-foot silver-and-scarlet opah or moonfish (*Lampris luna*), I wished I could have had my children with me and stayed in Japan for a year.

Two years after my visit to Japan, a telegram arrived at the Cape Haze Marine Lab informing me that the Crown Prince was passing through Miami for one day, on his way to visit South America, and would like to have me visit him. Unfortunately, I was scheduled to give a lecture in New York that day and had to express my regrets to the Japanese Council-General in New Orleans. A few weeks later, I had word from the Crown Prince again. On his way back from South America, he would stop in Miami for less than a day but would like to see me. I made plane reservations to be there.

My appointment to see the Crown Prince was at 8 P.M., shortly after he and his wife arrived in Miami. I thought he would be exhausted from his tour of South America, which I had followed in the newspapers, but he was sprightly when he received me. I was delighted to be able also to meet the beautiful and gracious Princess. The Council-General from New Orleans, the Master of Ceremonies of the Palace, and a number of other Japanese men were in the Prince's entourage.

At first they all appeared to be fascinated with our discussion of fishes, but one by one they began to look sleepy. I learned they had a tight schedule the next morning. The President of the University of Miami was going to escort the Crown Prince through the Marine Institute at 9 A.M. After this visit to one of the largest and most important research and teaching centers in our country, the Crown Prince was to visit the Miami Seaquarium at 10 A.M. Then his plane was to leave Miami for Tokyo at 11:20 A.M.

In consideration of this, I made the forbidden gesture to leave. "You must be very tired from your trip," I commented in sincere sympathy, pulling myself to the front of my armchair. "No, I don't tire easily," the Prince assured me, but he thought the Princess should go to bed, and she left expressing a wish that I join them on the tour through the Seaquarium in the morning.

And so the Prince and I continued to talk about fishes— the

taxonomic problems of Red Sea gobies and the clavicular apparatus of gobies in general, which is a favorite topic of the Prince. The other men in the room were sinking deeper into their chairs. The man sitting behind the Prince's chair had loosened his collar slightly (they all wore suits and ties) and his head had collapsed on his chest. The Council-General was making an all-out effort to keep his eyes from closing. He seemed to be sitting numbly in his chair, unable any longer to make the feeblest "How interesting!" comment. The Prince continued to discuss fishes with me as intently as when we had started.

When it was nearing midnight, I made another gesture to leave. The Prince assured me he wasn't tired; on the contrary, he asked if I would teach him to skindive! There was no doubt that he was earnest, and I even guessed he meant while we were in Miami. "When?" I asked. "We could go about 5 A.M.," he said. "Do you know where we could collect the goby *Bathygobius soporator*? I'd like to take some back to Japan with me." The Council-General's mouth as well as his eyes popped open.

I had no diving equipment with me, although I always carry a swimsuit. The Prince had no equipment either, but he had bathing trunks. If it wasn't for my good friend, diving ichthyologist and goby expert Dr. Dick Robbins, a professor at the Marine Institute who doesn't mind an unusual call for help at midnight, it couldn't have been worked out.

Dick brought along a graduate student, who was studying gobies, and all the gear we needed. I won't go into the security guard that the Japanese Consulate and the United States government, even on very short notice, give a visiting Crown Prince who wants to collect gobies a few miles from his hotel, but we had five cars with us. We all met in front of the hotel at 5 A.M., in the still black of night.

The first gray of morning was just lighting the horizon when we got to the stretch of beach near Virginia Key which leads to the mangroves and the habitat of *Bathygobius soporator*. We parked the cars at the road and walked along the beach with our gear, including a small mesh seine. All the Japanese and the American security guards were dressed in their suits and ties. Dick Robbins, his student, and I were barefooted and wore sport shirts over our bathing suits.

We came to an inlet where a fresh-water stream ran into the sea. We had to cross it, and the men who had to took off their shoes and socks and rolled up their trousers to wade across. As they did, the Prince spotted some fish and, with the Chamberlain and the Master of Ceremonies of the Palace, seined the stream several times and was delighted to catch some sailfin mollies.

The sky was turning pink and we could see the silhouette of the mangrove trees ahead. It was beautiful, calm and peaceful. Aside from our troupe, not another person was on the beach, which would be crowded with bathers and picnickers later in the day. Only one thing spoiled the scene. As the sky grew brighter, we could see that the whole beach was littered with beer cans! And as we waded into the water, the beer cans continued out to sea, littering the sand and mud bottom all around the roots of the mangrove trees. I felt ashamed of the careless habits of so many of our picnickers and wished that the trash collectors could have got here before we did. But the Prince didn't seem to mind in the least, He discarded his suit, shirt, and tie, down to the bathing trunks he wore, got a hand net, and waded into the shallow water to look for *Bathygobius*. Everyone joined the search. We collected a variety of fishes, Dick and I already diving with face masks. Then I came toward shore to see what luck the waders were having. They were walking around, ankle- to calf-deep in water, peering among the beer cans and mangrove roots. "Hey, what does this goby thing look like?" the Chief of Police, with a net in his hand, asked me.

"It's a small fish, about two inches long, sometimes nearly black, sometimes nearly white, depending on its background —but mostly it has dark bands across its back. It has a little suction cup on its belly." I waded away looking for *Bathygobius* before he asked the "why" question.

Then I saw a beer can partly sunk in the mud in a few inches of clear water. A small head was sticking out of the triangular punched hole. I picked the can up and emptied the contents into my hand net, and a *Bathygobius soporator* came flopping out.

We collected over a dozen of these gobies out of beer cans in shallow water. The early morning went by so quickly that there was only a short time left to teach the Prince how to spit in his mask to keep it clear and to get the mask and snorkel on him for a swim into deeper water while the patrol on shore watched us, in their rolled-up trousers, their shirts and ties, with worried faces.

Later, at the Seaquarium, the Princess told me, "My husband enjoyed the successful trip this morning so much. He loves to collect

To my left, the Crown Prince of Japan, Akihito, now Emperor. The Prince's Chamberlain, Meguro Katsusuke, on my right, Dr. Dick Robbins, University of Miami and one of his students in back.

fish, especially gobies, and there are so few opportunities for him to do it." She seemed to genuinely understand his interest, and I realized how difficult it would have been to arrange such a trip in advance. Considering the pomp and ceremony, the photographers and reporters, that accompanied all the planned events of the Crown Prince's visit to the Western Hemisphere, I finally understood why he asked so late to collect gobies and dive at 5 A.M. the next day. They both thanked me warmly as we shook hands, and the Prince and Princess and their group of escorts from Japan left the Seaquarium to make their 11:20 A.M. flight home with our collection of live *B. Soporator* for Prince Akihito's Palace Laboratory.

*Dr. Lester Aronson and his son Freddie who trained
the very bright young nurse shark that went to Japan.*

*Genie Clark with her shark tooth
necklace, when she was a professor
in the Department of Zoology at the
University of Maryland.*

Photo credit Mickey Irwin.

15 Mantas and Other Rays

GOBIES ARE THE SMALLEST vertebrates. One species gets to be an adult when it is half an inch long. It can be exciting for an ichthyologist to collect black and white gobies in beer cans or gaudily colored gobies in coral reefs, but there is a different excitement to collecting monstrous fish out at sea. Here most ichthyologists, especially women may "supervise," but really we let the experienced fishermen take over.

Our shark-fishing activities continued regularly and became almost routine. I seldom went out with the boat any more. The Lab's competent collectors didn't need me, and there was always so much work (especially on my desk) in the Lab that needed my attention.

But there were other big fish around that we couldn't catch on our shark lines, and when rare opportunities arose to observe or catch them—to heck with the desk work!

One day, when Dr. Wolfgang Klausewitz from Frankfurt was working at the Lab, I got a late start—missed the whole morning's work in fact—because I'd invited guests for breakfast. Dr. Adam Ben-Tuvia, ichthyologist, and Director Otto Oren, oceanographer, from the Sea Fisheries Research Station in Haifa, Israel, had come to visit the Lab. They arrived the night before, and I wanted to show off the spectacular view of Point of Rocks from our house. As we were eating breakfast on the porch, we saw the calm sea surface off the rocks start to churn. We rushed out and, through the exceptionally clear water, saw an immense school of full-sized tarpon. They were rolling at the surface in water only a few feet deep. I phoned the Lab and told John Strong to bring Dr. Klausewitz and cameras by boat. Adam, Otto, and I watched from the rocks until we could no longer stand it, got into suits and face masks, and then slid gently off the rocks into the school of tarpon. I had never been swimming with such a dense mass of large fish before. I didn't want to kill or collect one. I just wanted to experience the sensation of being in such a school. But we were outsiders, of course. Although an occasional tarpon came near enough to touch, most stayed more than five feet away (compared with the distance of less than two feet that they kept between themselves), and the school parted, then closed again as we swam through, taking an isolated area along with us. We could see John bringing *Rhincodon*, with Wolfgang Klausewitz rushing around and hanging over the railing. He had never seen tarpon like this, either. The school stretched for hundreds of yards. The *Rhincodon* anchored beside the three of us in the water.

"Why didn't you tell me you were diving with them!" complained Wolfgang, who was an experienced diver and had been the diving ichthyologist on Hans Haas's Xarifa expedition in the Red Sea and the Indian Ocean. Wolfgang became more and more envious as he saw the tarpon swimming around us and listened to our exclamations when we surfaced from a dive. By this time, many residents along the beach had spotted the school and had come to the edge of the water for a closer look.

John found a face mask on the boat and offered it to Wolfgang. "But—but—I have no bathing attire," he moaned. Finally, he could resist no longer. For Florida beach folks it was nothing too unusual, but for this straitlaced German Curator of Fishes of the Senckenberg Museum, I'll bet it's the only time he ever stripped down to his underwear in public before he leaped into the water.

Perhaps the largest fish in the waters of the west coast of Florida, besides the rare whale shark, is the manta ray. Like the whale shark, the manta is a harmless plankton feeder; only its great size can possibly be of danger to man, and then only to the man who invites the danger.

Most rays—and there are over 200 kinds—live on the bottom of the sea and have large poisonous spines on their whiplike tails to protect them from enemies bearing down on them, Occasionally these spines are used on a bather who happens to step on a flat sting ray camouflaged on the sandy or muddy bottom where rays can come in very close to shore. Most rays have their mouth located on the underside of the body behind a burrowing nose so they can pick up food as they swim along the bottom. They have eyes on the uppermost part of the head for the best view from the sea floor. And since they often lie so flat on the bottom that they can't use their regular gill openings located on their underside for breathing, they take water in and out of a large pair of spiracles—holes in the top of the head behind their eyes. Their gills take on other interesting functions. By holding the body flat as a pancake then sucking in its center gill and belly region, a sting ray can create a vacuum to hold its body tight against the bottom. By taking water in through the dorsal spiracles and releasing it through the ventral gill slits, a ray can control a slow glide or sudden jet-like takeoff from the bottom with little use of its "wings," which then gracefully flap or undulate and propel the ray once it has taken off for a swim.

But the manta ray is a ray that has become adapted secondarily for life near the surface of the sea. Its wings are enormously developed for fast and constant locomotion. Its tail is short, and only a blunt little knob remains in place of the formidable spine of its ancestors. The manta's mouth is at the front of its body and has lost any traces of the unique plate-like crushing teeth used on clams and other mollusks (the food of sting rays). Instead of a projecting snout, the manta (or devilfish) has a pair of "horns" at the front of its head that it can point forward in the streamlined shape of a bull's horn or unroll into flattened scoops under its greatly enlarged mouth. Its mouth is an oversized colander which filters plankton from the water. The filtered water passes over its internal gills and out of its openings. It lacks spiracles, and its bulging eyes are located on the sides of its head.

Viewed from shore, mantas cause a commotion—not because they are such odd-looking and spectacular animals but because they are usually not seen except for the tips of their wing-like pectoral fins, which often protrude, together or one at a time, as the manta glides near the surface.

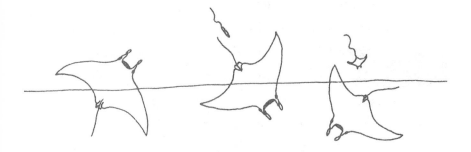

MANTA GIVING BIRTH ABOVE WATER

MANTA FEEDING

(AFTER RUSSELL J. COLES, 1916.)

People would call the Lab and report "a whole school of sharks," stating firmly that they knew these weren't porpoises or tarpons because their fins didn't "roll" at the surface. When we'd learn that some of these "sharks" were swimming in pairs because the observers saw two fins traveling together, we knew the mantas were back again.

Why mantas come close to shore in groups at some times of the year is not known for certain, but several interesting reports based on fascinating observations give some clues. Dr. Russell Coles reported early observations on the behavior of mantas up to 25 feet wide and a smaller, 4-foot-wide, horned relative, *Mobula*. He claims that *Mobula* is the "devilfish" that leaps gracefully out of the water, sometimes clearing the surface by more than five feet and returning to it with a loud noise. The manta, Dr. Coles is certain, does not leap clear of the water but rushes head first upward, until approximately half its body is out of the water, and then it rotates like a wheel on its axle. He made a remarkable observation (see figure) of a harpooned manta giving birth during a leap, the single embryo shooting out like a cylinder, unfolding its pectoral fins, and descending like a bird.

A group of Italian divers reported seeing "the dance of the mantas" at dusk in the Red Sea. The mantas floated to the surface, turned over backward, and dived down, continuing this acrobatic over and over for nearly an hour. The divers saw newborn mantas in the water, and on closer looks they could see tails of the yet unborn embryos protruding from the *cloaca* of the large females still rolling over. The observers concluded that this acrobatic is part of parturition. Other divers have called this behavior a "mating dance." We scientists interested in the sexual behavior of fishes can be exasperated by the incongruous modesty of some French and Italian divers who claim to have witnessed mating orgies of mantas but are shy about reporting details. Stan Waterman made a magnificent movie of mantas "looping" in Tahiti, and they were not giving birth or mating at the time.

The central Gulf coast of Florida had numerous mantas at one time. But now—like the trunkback, the largest of all turtles and a species once common in these waters—mantas are becoming rare. Theodore Roosevelt, during the manta season, came to Boca Grande and caught one. He wrote an article on his thrilling experience.

A plane pilot phoned the Lab one morning to tell us that he had just seen six large manta rays only half a mile offshore of Sarasota. The Gulf was smooth, a perfect day to hunt them. John Strong phoned his friend Dick Miller, a local sports fisherman who had quite a bit of experience harpooning large rays. The *Rhincodon* wasn't ready so we took the *Dancer*.

248

It was in November of 1961, rather an off season for spotting mantas in the area. Since 1955, our records of manta sightings previously had been in spring, summer, and early fall.

It was noon when we located the protruding pectoral fin of a basking manta and approached the creature cautiously. It looked like a gigantic black blanket spread on the sea. We felt disappointment and heightened excitement as we got a good look at it. We wanted a manta, but the size of this one made us pause in our planned attack. It was larger than we could or should try to capture. Six-foot-three John Strong and six-foot-nine Dick Miller suddenly looked small compared to the enormous fish swimming lazily around us. Its pectoral fins were white-tipped and when it lifted one of these wing-like fins out of the water as it banked and turned, we got a glimpse of the all-white underside. The rear corner of the fish tapered to a small tail, and the fourth corner, leading, had two horns and sieved the sea with a huge mouth, forever open, between the horns.

John and I exchanged glances and knew what we were both thinking. We would undoubtedly lose an expensive spear and perhaps 150 feet of 3/8-inch nylon line and our yellow buoy, but we also knew we couldn't resist a try.

Dick steered the boat and turned off the motor as our boat was almost over the manta's tail. John was ready with the harpoon gun and fired into several tons of flesh. The manta exploded into action with an immense splash and then sounded, pulling the line and buoy after him. Dick started up the motor and we went after the streaking yellow buoy.

John had aimed at the center of the manta, not one of the huge muscular "wings," as he would have if we wanted to keep it alive, but it showed no signs of being hit in a vital area. It was twenty minutes before the manta slowed down sufficiently for us to catch up with the buoy. Once we secured the buoy and end of the line, we turned off the motor again. Our boat kept moving almost as fast as before. The silent glide of the boat felt very strange. The nylon line disappeared into the water in front of the boat. We couldn't see the manta; it stayed deep and pulled us steadily.

We were less than 300 yards from the beach off Longboat Key, Sarasota, when John speared the manta. We were now much farther out, but heading in a southerly direction, and I hoped the manta might tow us to the Cape Haze Marine Laboratory on the south end of Siesta Key, going in close to shore near Point of Rocks so that I could blow the horn as the Lab's boat passed my house, and our nursemaid Geri, with Niki and the three other children, soon due home from the Out-of-Door School, might see us being towed by the huge manta.

The manta towing our boat now showed no trail of blood, no sign of weakening, and unobligingly started heading directly out to sea. It pulled us steadily and smoothly, as if we were attached to a submarine on a well-defined straight course rather than to a wild injured creature. Was this manta fleeing in a state of panic, without bucking or struggling to try to shake loose the harpoon, or did it only feel the first surprise hit and flee from the spot where its back was tapped, unaware that it was bound to its pursuers?

After over an hour of only slightly altering its course a few times, it finally surfaced and we again saw its great body. We dropped anchor, but the manta's pulling force seemed unaffected. By now we were far from shore, and the hopelessness of the situation was sinking in as the sun got low. Then suddenly our hopes were raised as a larger boat came speeding toward us. It was Dr. Douglas Williamson, an ophthalmologist who, when he wasn't looking into patients' eyes, could usually be found a few miles out in the Gulf skin diving. An airplane had seen our plight and radioed Doug, who wanted to join in and help us get the manta. A number of people now knew from the plane pilot's message that we were hooked on to a very large manta.

Doug's 33-foot boat came alongside, and he and his crew were amazed at the speed the manta was towing us with our anchor down. Doug threw us a line, but we didn't notice any decrease in the 3-to 4-knot speed even with the additional drag of Doug's larger boat. He turned off his motor and threw over his anchor, and still we headed straight out into the Gulf of Mexico at a fast clip.

The bottom in this area is mostly sand, but we knew that seven miles from shore there began some limestone reefs, covered with mollusks and myriads of invertebrates that glued the soft limestone together into harder reefs. I'd gone diving on these reefs with Doug to see how his explosive underwater spear gun worked on large groupers and sharks. As we approached the reefs, I imagined the jolt we'd receive when the anchor caught, and visualized the snap of the line as the manta became free.

The tallest buildings in Sarasota were now only spots on the eastern horizon and the sun was getting ready to set opposite, but we couldn't cut the line. We were thrilled by the silent tandem movement of our boats. With the bright sunlight and shoreline almost gone, the mystery, danger, and exciting wonder of the sea grew.

Finally, I decided it was nonsense to continue. The manta was too strong for us, and it wouldn't surface where we could attempt to spear it again. "We'll have to cut the line," I called to Doug. He was crushed with disappointment. "I haven't even seen it!" he protested.

Then he got what he thought was a bright idea and I thought was a foolhardy one. He had his explosive spear gun aboard, and wanted to slide down the harpoon line and blast the manta in a vital spot. There was still enough light to see a few feet underwater and he needed to get only close enough to bump the end of his long spear gun on the right spot. I wanted no responsibility for any part of this idea but I couldn't dissuade him.

"You're the one who told me mantas don't have poisonous stings on their tails," Doug argued.

"But the size of this one—one flip of his wing tip could knock you unconscious," I said.

However, Doug reasoned that if the harpoon now in the manta was close to the center of the body (as we knew it was), by going down the line, he would be at the center of the creature—out of danger, he supposed, of the lashings of wings or tail. There was no stopping Doug.

As we watched him going down the line, his scuba gear over his black wet suit, and saw him disappear into the now dark gray water, I had misgivings. As Director of the Cape Haze Marine Laboratory and this operation, I should have stopped him. All I had to do was tell John to cut our line.

Now it was too late. We heard the explosion. Doug had hit the manta. The boats suddenly slowed and the line slacked, and we all strained our eyes to see down into the dark water. Suddenly, Doug popped up screaming that the manta had turned and was following him. "It's the biggest thing I've ever seen. It came after me with its horns!"

Doug climbed hurriedly back into his boat. There was splashing and I saw a little blood in the water, but none of us got a glimpse of the manta. The slack line became taut again and, the boats were going in circles. The manta was turning and pulling us this way and that. Its movements had definitely changed, and we hoped it had been seriously injured.

"Where did you hit him?" I asked Doug.

"Right in the skull."

"What 'skull'?"

Rays, like sharks, have no bones and no true skull, but more than that, a manta doesn't even have a head that you can recognize. Behind the "horns," it just widens out to the wing tips. It turned out that Doug had exploded the spearhead about a foot behind the front end in a center line between the two horns, where he imagined the manta's brain was located and where it is likely to be in any large fish. What Doug didn't realize is that because the manta feeds on plankton and has such a big oral strainer, the whole central front end of the manta is mostly mouth. Probably only the roof of the mouth had been hit—a nonvascular area hardly a place to weaken the manta. The sun set. The manta seemed irritated, judging from the erratic way it now pulled the boats.

Then came a sudden stillness; the boats stopped moving. "We've lost him," John said as he pulled up the limp harpoon line and finally the twisted heavy metal spear. Everything came to a dead halt, and we looked at each other with great disappointment.

Then, before any of us spoke, something unbelievable happened. *We were moving.* No one had turned on the motors, but once again our boats were moving in tandem out to sea, faster than ever before. We couldn't understand it. "He's got the anchor line!" John gasped. And sure enough we could see the taut anchor line from our bow.

It took us awhile to realize what had happened. The old- time sponge divers, who used helmets when they dived, knew mantas didn't bite or have poisonous stings like other large rays. But they feared having the lifeline from their helmet to the boat caught between the horns of a manta. This is exactly what had happened to our anchor line, and the manta was pulling us with the anchor in its mouth.

We all cheered and shouted comments about the wild ride we were having. But it was soon over. The manta, turning abruptly, gave us a short good run for that twisted spear until, on one turn, the anchor fell out of his mouth and he left us for good.

The next day we went out again. I had told the children about our adventure, and Tak developed a mysterious stomach ache in the morning, which disappeared as soon as the school bus took Hera and Aya. So Tak joined us on the trip. And he will never let me forget that he was the first to spot a manta that day. With the help of another boat and Jack Gilbert, Joe Cash, and John and Dick—all throwing harpoons—we finally slowed down a much smaller but still sizable manta, which Joe could then shoot with his .32-caliber rifle.

The Sarasota City Pier was packed with people who had heard from radio contact that we were bringing in a manta. My secretary, Mary Eastridge, came to double as photographer and take dictation during the dissection. Kay von Schmidt, our research assistant, also came from the Lab and made a stop at school to pull out her two daughters and my two. And Geri brought Niki from home.

It was a tremendous job to lift the manta onto the pier. It weighed 2,200 pounds with part of its body not lifted free from the ground. It kept sagging. Its awkward body and lack of a skeleton made it hard to hoist by rope. (I later wondered if they ever really weighed, or just estimated, the 5,000-pound, 19-foot-6-inch-wide manta I learned Captain Frank Roberts, Sr., brought to the same pier in 1936.)

Kay and I dissected the manta on the pier, and I ended up giving a lecture on its anatomy to the crowd of spectators. The well-developed structure of the abdominal pores of this ray bore out a suspicion I had about the function of these structures, which were always better developed in sharks that swam about through various depths than sharks which spend their time on the bottom like nurse sharks. The pores seem to be part of a pressure-adjusting mechanism in those fishes without swim bladders.

The owner of the Oyster Bar restaurant, at the end of the pier, offered to cook some of the manta. The dissection took a long time, and deep-fried breaded manta was passed around to the audience as I cut into the pericardial chamber and we watched the huge heart, the size of a baby's head, still beating.

Lionel Murphy is a flying photographer. He does much of his photography taking aerial pictures through the open side of a small plane with the door removed while Pilot Harry Louden banks the plane so Lionel can shoot straight down. Over the years, no one has gotten a better view of the changing scene in Sarasota and the central west coast of Florida. The wild mangrove forests, full of birds and wildlife, are gradually being cleared away and replaced by neatly trimmed lawns with concrete seawalls abruptly edging the area where once the land sloped gradually among mangrove roots through an inter-tidal zone of myriad forms of marine life. The panorama of aerial photography has recorded, better than any words or cries of fishermen, the dwindling fishing grounds of our once-rich bays as the multi-fingered hands of developments keep reaching farther into the decreasing expanse of water. Dredging and filling operations keep covering and killing the grass beds and the nurseries for microscopic life, crabs, shrimp, and fish. The bays are becoming more polluted each year, and in many places, local law must now prohibit the taking of shellfish for human consumption. The sandy Gulf beaches on the outer sides of the keys were once dense with tangled sea-grape trees, behind which large oaks, laden with wild orchids and Spanish moss, hung over poison ivy patches where pumas and snakes could find constant cover. Now the lawns even reach the sand in many places, and in these flattened clearings regularly placed coconut palms, whose stunted form, brown-edged leaves, and shriveled milk- less fruit struggle to stay alive just outside their natural range, give the illusion of the tropics to a semitropical area that is touched by frost almost every winter.

Lionel Murphy also recorded an extraordinary biological phenomenon from a flight he and Harry made on July 18, 1959. Several years later, it occurred to him to visit the Cape Haze Marine Laboratory and show me the photograph he had taken. As he unrolled the 15- by 20-inch enlargement print on my desk, I was utterly amazed.

It was a school of rays such as I would never have thought could exist, least of all in Sarasota on the shallow sandbar in Big Pass! Lionel had taken the photo from about 300 feet above the water on an afternoon when the light was perfect and the water was clear. There seemed to be over a thousand rays in several layers in the dense school. How could such a phenomenon occur in water next to shore and next to the traffic of all the boats that go in and out of Big Pass without causing a stir or anyone calling the Lab about it? Yet this had occurred several years before and no one had reported it to us.

Lionel was pleased that his photograph caused so much excitement at the Lab. "I can make all the prints you need of this picture," he assured me when I asked if I could borrow the print.

That evening I couldn't eat my supper until I had made a count of the rays that were distinct individuals in the upper layer of the school. There were 3,050. And there seemed to be at least two thousand or possibly three thousand more rays in the overlapping underlayers.

At first I thought from their protruding, rather pointed heads that they were spotted eagle rays (*Aetobatus narinari*). These are among the more common large rays around Sarasota, and in 1910 Dr. Russell Coles reported a large school of them passed under a yacht near Cape Lookout, North Carolina. The school was about three feet below the surface.

But as I studied the photo with a magnifying lens and then Lionel made further enlargements of certain sections of the school, the protruding heads showed up as blunt. They were clearly cow-nosed rays (*Rhinoptera bonasus*). All the fish were pointed westward. We could figure this from the sunlight reflected by the ripples and the time of day, which Lionel calculated from his log was very close to 2:30 P.M.

Were these rays really schooling? Why should a large fish like the cow-nosed ray, with such a formidable protecting device as a whipping tail with dangerous and poisonous spines, need to gather in such a dense mass? What were they doing in Big Pass? Mating? Feeding? Cow-nosed rays are found all year round in Sarasota, so this didn't seem to be a gathering in preparation for some mass migration. We knew the fishermen avoided rays but sometimes caught groups of them in nets.

A photograph alone can give a misconception of true schooling behavior. For example, fish could appear to be schooling if, like a group of cows standing in the wind, they are crowded and all facing into the current. I checked the tides at Big Pass on the day the photo was taken. It was flowing out of Big Pass at 2:30 P.M., and therefore westward, in the direction the rays faced. Lionel remembered circling the rays for some time, and noted they didn't move from their position on the sandbar. It's remotely possible that the rays were facing into a countercurrent eddy over the sandbar or were in an area of relatively still water.

Big Pass is about 2,000 feet wide, with shifting shallow sandbars usually covering over half the width. The oval- shaped school of rays measured approximately 350 by 180 feet. The sandbar the rays were on is known to have, at various times, large beds of the clam-like bivalve *Chione*. Perhaps the rays, with their blunt snouts for burrowing in the sand, were feeding on an unusual crop of *Chione*. If only I could have examined the stomach contents of one of those rays! Or looked at the condition of the gonads to know if they were in mating condition. Did such huge schools of cow-nosed rays form often, perhaps unnoticed in deeper water, or did Lionel catch a rare happening?

I looked through my books and began asking all the sports and commercial seamen and plane pilots I could find in Sarasota if they knew about big schools of rays, and I advertised in the newspaper and the Lab's newsletter for further information. Yes, large schools had been seen, but from boats or bridges, not planes, and the observers (and fishermen are not usually conservative) estimated the rays were in the hundreds. The descriptions (plain brown, protruding but blunt head, about 3 to 4 feet wide) all fitted the cow-nosed rays.

Jay Odell of Venice saw a school half a mile offshore about sixteen miles south of Big Pass in August, 1956. Fred Logan, who had taken the wonderful *Serranus* movies for me, told me he and his wife had seen at least six large schools in the passes in this area in the past ten years, in spring and summer. And several other boating residents and fishermen reported seeing the schools in Sarasota Bay itself in spring, summer, and fall. The schools were known to stay in one area for days, and sometimes up to two weeks under the bridge near Byrd's Fish Camp at New Pass.

On June 24, 1964, a day I set aside to tackle the growing unanswered-mail problem, the Coast Guard called to report that one of their helicopter pilots had just spotted the rays forming massive schools off nearby Anna Maria Island.

A few phone calls to Washington, including one to Dr. Sidney Galler at the Office of Naval Research and a long- distance plea to Admiral Stevens in Miami in charge of the Coast Guard District, and I got permission to use a Coast Guard helicopter to make observations and movies. We took off from the St. Petersburg Coast Guard Station, where first they bundled me into an oversized space suit and helmet with an intercom system (I never could make out what the scratchy voices said to me). Lionel Murphy, the experienced aerial photographer, sat snuggly in the cockpit taking color photographs from an open window, and they put me, tethered by a loose cord hitched to the back of my suit, in the prize position for observation. They had removed the right side of the helicopter and I sat on the floor with my legs dangling out of the plane, clutching the Lab's Bolex movie camera while the pilot banked the plane so I could shoot straight down and take movies of more than 20,000 cow-nosed rays massing near the shore of Anna Maria Island. I felt like a very strange and scared marine biologist.

The *Rhincodon*, with every strong arm and spear we could round up on short notice, joined us, and I photographed it going through the schools of rays (sometimes five layers deep) as we finally got the specimens we wanted.

The summer and early fall of 1964 and 1965, we took movies and made studies of schooling rays from Anna Maria Island down to Manasota Key. It was not a rare phenomenon. We studied over a dozen such occurrences. Our summer students (and my children, of course) took turns riding the *Rhincodon* and spearing rays. Sometimes we went in Harry Louden's plane, sometimes a helicopter. Sometimes the water visibility was poor, the surface too rough, the time of day wrong for proper light. From the plane, we could direct the *Rhincodon* by radio to the exact place of the school, but we started to understand why these masses of rays were rarely seen. The *Rhincodon* could be less than fifty feet away from the school, yet unless the conditions were ideal the people on the boat couldn't see the rays, which stayed under the surface and, unlike the manta, seldom let a tip of their pectoral "wings" protrude into the air.

In spite of dissections—and Kay and I dissected over a hundred rays—we could not find out why such huge schools sometimes formed. Kay von Schmidt was the most valuable research worker we ever had at the Lab. She learned to dissect sharks and rays better than I, and with more persistence. Once when some fishermen accidentally tangled their nets in a big school of rays on a sandbar, they had to bring in the whole load to dislocate the rays, which had their jagged stingers caught and twisted in the net. Kay learned where the fishermen dumped the rays, and with a crew of volunteers, plus some drafted helpers such as our children (who by now had seen us dissect enough cow-nosed rays for a lifetime), we went to the fish camp and Kay pulled up the rays. The rays were fast decaying in the warm water, but Kay continued taking measurements and making dissections long after I was ready to give up.

We learned a lot of negative answers to our questions. The stomachs of rays from the various large schools we studied were mainly empty. No great *Chione* feast was going on. The gonads showed it was not the mating season. Many of the female rays were carrying embryos in various stages of development but none near ready for birth. So it was not the phenomenon ascribed by some authors to a massing of manta-ray females to give birth to their young in one place.

The schools of cow-nosed rays were a mixture of adult males and females, and even some immature, though almost fully grown, individuals.

Some of the schools I photographed from the air moved slowly along the shore, changing shape like an amoeba. Sometimes one was ovoid, like a giant ray itself, stopping to turn into a horseshoe, then stretching out and winding along the shore like a giant snake. Sometimes two schools, each with thousands of rays, met and fused and traveled on as one. Occasionally a school would split up into two or three schools and then reform as one.

What currents and other conditions in the environment or within the rays themselves controlled this massive gathering movement? If such schools can exist practically unknown along populated resort beaches, what schools and other massive groupings of animals may take place in more secretive parts of the sea completely unknown to man? I would like to know if and why. To others the significance may be greater. Could a man (and of what nation?) learn to control such massive schools to acoustically camouflage a submarine, the way the elongate trumpet fish, which ordinarily hides still and vertical in plant-like soft corals, takes along a school of yellow surgeonfishes as perfect camouflage when it travels horizontally away from the reef? What symbiotic relationship will man have with fish as we continue to live longer and deeper in the fish's environment?

Kay vonSchmidt, co-author of "Sharks of the West Coast of Florida," one of the most valuable research workers the Lab ever had.

16 The Laboratory Grows

THE CAPE HAZE MARINE LABORATORY kept growing larger, externally and internally. It matured well with experience and became a place respected and noted for marine biological research. But in this evolution the necessary changes occurred, as when a simply delightful baby turns into a complex adolescent. For me it meant richer and deeper pleasures, occasional disappointments, and increased responsibilities. We had a good, but not entirely true, image to live up to and to try to correct. The accomplishments of the Lab, I felt, were truly valuable and some of our findings wonderful. We were getting the over-all picture of the local marine life, the interrelationships of the biota with the entire environment. But piles of data, analyses, and reports forming the basic part of the local marine biological picture were taken for granted from the beginning. When the Lab was less than a year old, we were considered the authorities on the local marine life. The shark-training experiments were spectacular and glamorous and got publicity and acclaim out of proportion to their real value. The fact that a woman was conducting studies on such fearsome fish increased the attention. Only experienced biologists who had worked in the field came to the Lab with true understanding for what we were learning and what we still didn't know.

Our study of the local marine life is far from complete. Probably such a study never can be complete. Many problems remain unsolved; many important questions still need answers. We haven't yet solved the mystery of the shark's abdominal pores; we can't predict, prevent, or alleviate the damage of the "red tide" to marine life. And when we have the knowledge of how valuable our shallow bays are, we can't prevent our fellow men from dredging and filling and damaging the marine life without repeated court battles and a struggle on the local economic and political score. Having watched the slow and carefully planned development of the Cape Haze Corporation and knowing the Vanderbilts' concern about the quality of their development— what it would do for the community, how we could learn more about and make the most of our marine life—it was hard to accept what was going on elsewhere. The "fast-buck" developers, especially those wantonly destroying the bays, luring people to buy "waterfront" land where the water was being turned sterile, was like watching the slow murder of the goose that laid the golden eggs. To those of us who knew, statements such as "Dredging and filling is good for the bays" sounded as sensible as if a tobacco company, faced with irrefutable fact, claimed that "Cancer is good for you." Fortunately, a businessman who stands to make thousands or millions of dollars by filling a bay may think twice and make the right decision, because he has felt the magic of the written word of Rachel Carson or the spoken word and the charisma of Jacques Cousteau.

Other frustrations, such as trying to prove to an unbelieving group of scientists that Bill Royal had indeed found a human brain and ancient human skeletons thousands of years old, the oldest dated in the Western Hemisphere, were finally resolved after years of slowly accumulating enough convincing evidence and support.

We found no coelacanth, no Loch Ness monster, no shark repellent. Although our sharks could tell different-colored targets apart, we could not prove they had color vision. But we had many satisfactions and fulfillments, no small part of which were the pleasures we had from the hundreds of visiting scientists and students who used the Lab as a base for collecting and studying marine life and the teamwork that resulted. Every interested fisherman and nature lover (which I basically am too) whom we encountered through the Lab's name contributed to the drive and stimulation to go on studying marine life.

From visiting scientists I learned about other studies, techniques, and outlooks. Our students, boatmen, collectors, lab and field assistants, and secretaries have each caused some renewed self-examination of my work and an appreciation and enjoyment of life I could never have gained by working alone.

The chapter of the Cape Haze Marine Laboratory—perhaps the best chapter of my life—now belongs to the past. As the Laboratory changed and grew, my personal life changed too. My marriage to Ilias came to an end, and I decided to move north with my children.

Among the many struggles I had with my conscience was what to do about the Cape Haze Marine Laboratory. I suddenly felt that I was abandoning a child who still needed me.

There was one person ideally suited to take over—someone with superb administrative ability who could handle with ease the metamorphosis of the Lab from a field station to a major research center. Perry Gilbert, an outstanding zoologist, teacher, and lecturer, has the imagination and ability to create and carry through research programs in the plushest laboratories or the most primitive field conditions. He heads the Shark Research Panel of the American Institute of Biological Sciences and leads worldwide studies on sharks. He had followed the developments at the Lab from the beginning and often said, "Genie, you have a wonderful setup."

A major factor would be to find enough additional support for a director's salary, since I'd given up mine, and for the kind of burgeoning research programs that would start if we could obtain a director such as Perry Gilbert. The Vanderbilts had already given much more support to the Lab than they originally had in mind when we started as a small field station and were heavily committed to helping many other organizations. As their children grew up and were in or approaching college age, the Vanderbilts decided to settle farther north.

While I was wondering what would become of the Lab, an unusual man stepped into the picture. Mr. William R. Mote was in the transportation business and wanted to retire early and use his earnings and energy to do something worth while connected with the sea he loved so much. His hobby was fishing—casting for a snook near mangrove roots in shallow bays along the west coast of Florida, the area where he was born and raised, or hooking onto big-game fish all over the world, such as his record black marlin of 1,180 pounds in the deep water off Cabo Blanco in 1957.

His interest in fishing brought him into contact with marine biologists and oceanographers, further whetting his appetite for ocean study. He was a member of the Finance Committee of the Lerner Marine Laboratory and was serving on the Florida Council of 100's Committee of Oceanography. He decided he wanted to further the study of the sea by building a marine laboratory on the west coast of Florida. He came to me for advice on how to go about it and wanted to be sure such a venture would not interfere with our laboratory, and hoped his laboratory could cooperate with and complement the kind of research being done at the Cape Haze Marine Laboratory.

The developments in the next few years were gradual, carefully planned, and at times complex, but the Cape Haze Marine Laboratory evolved into the Mote Marine Laboratory. Bill Mote retired and moved to Sarasota from New York, and his interest in helping promote more knowledge of the sea, through the medium of the Lab, became his major preoccupation. Perry Gilbert became the Director in 1967 and could maintain his association with Cornell University, which became affiliated with the Laboratory. Bill Mote turned over to the Lab 30 acres of his land, coincidentally located almost next to the original site of the little wooden building where Beryl and I started the Lab in Placida in January, 1955. He also gave the Lab a 27-acre unspoiled mangrove island, Devilfish Key, in Charlotte Harbor. The Charlotte Harbor additions to the Lab property will allow the Lab to expand in many directions. Architectural plans have been drawn for buildings with laboratory rooms to accommodate at least twenty investigators and their assistants at one time. Ponds on Devilfish

Key can be used for ecological and maricultural studies.

I am still a frequent visitor and a member of the Board of Directors of the Laboratory. Also as a Research Associate I can enjoy all the benefits of studying fishes there whenever I wish, watching the great progress being made in the expanded programs to survey the still-unpolluted Charlotte Harbor, and many other investigations on the marine life of the estuaries and the marvelous continental shelf of the central west coast of Florida and related areas. With relief and gratitude, I see the awesome and wonderful weight of the responsibility to carry out these enormous research programs on the strong shoulders of Perry Gilbert with his team of capable co-workers.

In a freedom I've never known before, I find myself beginning a somewhat new way of life. I feel refreshed, stymied, aging, and reborn with the daily problems of the charming and awful adolescents I live with and the challenging young adults who look to me to teach them ichthyology at the University of Maryland. What abdominal or abominable pore will perplex one of them? Which of my children or students will be diving with me next summer when we exchange a communication without words and exclamations of only tones and bubbles at some wondrous sight we shall see together on the bottom of the Red Sea? How many more marine laboratories will be born along the shores of the sea while we continue to wonder at and try to solve its mysteries?

William R. Mote and Perry Gilbert in front of Mote Marine Laboratory when it moved to City Island, Sarasota in 1978. In 1980, Mote's Public Aquarium was added onto this site where Bill and Perry are standing

Kumar Mahadevan, current and longest director of Mote Marine Laboratory, Bill Mote, benefactor, and Alfred Goldstein, 1998. Alfred Goldstein and his wife Ann, were major donors for the Aquarium, the Marine Mammal Center, and other projects at the Mote Marine Laboratory

17 Fishy Adventures Continue

TODAY I AM BACK at Mote Marine Laboratory happily ensconced in my second floor office in "Shark Alley" in the main lab building. As a volunteer, with such titles as Founding Director, Trustee, and Senior Research Scientist, I can come and go as I please and enjoy the camaraderie of many wonderful friends and colleagues that make up our "Mote family," and the many visitors from around the world who visit Mote. Elasmobranch scientists often stop by for a few days or do research here for months, and give seminars on their research.

I don't keep regular hours. I travel frequently on expeditions, give lectures, take part in conferences on fish or consult with colleagues in other parts of the world. At 89, when I'm home in Sarasota, doctor's appointments often have to take priority. But I do like to "work" (= play) at Mote every day I can. Sometimes I work late into the night or weekends to observe the nocturnal activities of our live adult convict fish, *Pholidichthys*. I can observe and photograph their strange nocturnal "tunnel-building" via a television monitor on my desk that is connected to the experimental fish tanks behind the public exhibits in our aquarium building. In the quiet of night, when only the night watchman comes to check on me, we can record from my office, undisturbed, this marvelous nocturnal activity. We now have a major exhibit, open to the public, which allows them to see the convict fish in their tunnels during the day, using mirrors.

After a short stint teaching at City University of New York, my children and I settled for 32 years, from 1968 to 2000, in Bethesda, Maryland. I became a Professor in the Department of Zoology at the University of Maryland. I taught six different courses in marine biology, wrote 12 articles for *National Geographic* along with the great photographers David Doubilet and Emory Kristof, married two more fascinating men, and gave up spearing fish. As a Professor Emerita, I still travel to Maryland to work on scientific manuscripts with my wonderful assistant and friend, Bev Rodgerson. She edits my writing, and plans and organizes our many diving expeditions to study the behavior of fishes in various parts of the world. Many details of our diving expeditions have also been planned by Martha Kiser and Joan Rabin. These expeditions and the resulting scientific publications have been sponsored mainly by the University of Maryland Foundation, the National Geographic Society, and many private individuals.

Over the years, I've dived with more than 50 whale sharks in Ningaloo Reef, Australia and near LaPaz, Mexico (the largest 50 ft.

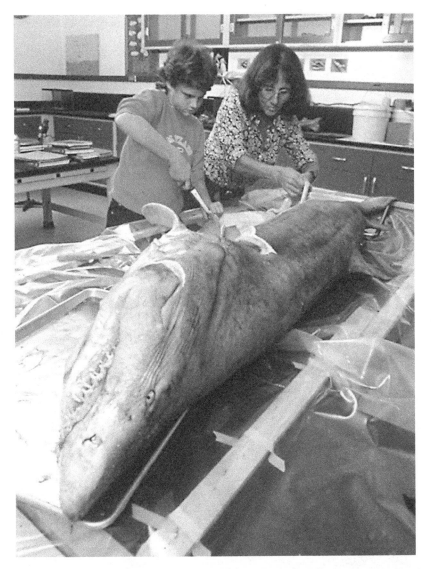

Dissecting a sand tiger shark in my lab at the University of Maryland with Jessica Rabin, age 9, the youngest student ever registered at U of Md.

long), dived with over 30 white sharks in south Australia (in cages from which I could reach out and pet them), been chief scientist in charge of 72 submersible dives to study deep sea sharks. In the

ALVIN we dove to 12,000 ft.; we saw the hooded octopus, but no sharks. In Japan, when diving to 140 feet off the Izu Peninsula, I was carried off by the largest crab in the world, which gets to 13 ft. across with its claws spread open. In Suruga Bay, we recorded the largest shark ever filmed in the deep sea, a sleeper shark, *Somniosus*, bigger even than our submersible. Once our submersible got hung up in a submarine canyon in Grand Cayman, and we wondered if it would get loose before we ran out of air. My most dangerous accident happened in my bathtub—I slipped and knocked myself out, thinking to myself "Shark Lady Dies in Bathtub!"

It was my dear late husband "Yoppe" (Yoshinobu Kon) who brought me back to Sarasota. He was born and lived his whole life

Thirty years later, and still messing around with the dissection and study of large sharks. A big eyed thresher shark is studied with Dr. José Castro in the Mote Marine necropsy lab.

photo credit José Castro.

in Hawaii. He and his first wife, Bertha, lived all their married life in the beautiful Manoa Valley of Honolulu. Yoppe was best friends for over 50 years with my first husband, Hideo "Roy" Umaki. Whenever my children and I went through Hawaii, we stayed with Yoppe and Bertha. After Bertha died, Yoppe and I decided to get married. However, he couldn't stand the thought of living in the snowy winter of Maryland. He suggested, "I'll give up Hawaii if you give up Maryland. We could live in Sarasota near your children and your grandson and the Mote Laboratory you love." In the next few years, we enjoyed the sunsets and warm sea air from our condominium balcony in Sarasota. Yoppe joined our expeditions to Papua New Guinea and the Solomon Islands. I gave in to his love for cruises on big ships to Alaska and other parts of the world, and discovered, to my surprise, that I could work on my manuscripts while he played duplicate bridge. He never resented my work and my frequent commutes to Maryland to close up my laboratory. He knew I was making the move to the warm climate he loved and the lab I loved, and indeed we got married in the Mote courtyard. He was happy taking it easy in our condo while resigned to the Parkinson's Disease that took him away from me after only three years of marriage. I carried his ashes to Hawaii to bury them beside Bertha whom we both loved so much.

Now, I enjoy opening the fun door (see photo) to my cluttered office with all my favorite books and reprints, and wall space filled with pictures of sharks and other fishes. Jaws from sharks we collected are mounted around the ceiling. Pictures of my ichthyological colleagues dominate the walls around my computer. My file cabinets and shelves are crammed with notes on fish studies from more than 100 expeditions abroad, mainly dive trips where I have studied fish

The door to my office is always open.

underwater. We have a fine Mote library just down the hall where our librarian, Sue Stover and her excellent volunteer staff can find the most obscure references I need for the scientific manuscripts I am writing. If they are not on the shelves, she finds them on the Internet. On the way to the elevator, just down the hall from Shark Alley, lives miracle man Henry Luciano who maintains nearly 500 computers in the Mote complex. I am spoiled by having him just seconds away from my office. Further down the hall, in view of the elevator, is the open door of Mote's wonderful President, Kumar Mahadevan, who often works later than I do and waves, "good night," when I leave. We all appreciate his leadership for the last 25 years.

It's hard to believe that 56 years ago, in January 1955, I moved to Florida from New York with my husband, Ilias and my two daughters. When we began, the Cape Haze Marine Laboratory was in a little wooden building 20' x 40' in the fishing town of Placida on Lemon Bay. Today it is the Mote Marine Laboratory with headquarters in Sarasota, and field stations in Summerland Key, where they study coral reefs, in Charlotte Harbor, where they study the largest unpolluted estuary on the east coast of the United States, and the 200 acre Mote Aquaculture Park, east of Sarasota. Our main buildings on Sarasota's City Island include the Mote Marine Research Laboratory, Ann & Alfred Goldstein Aquarium, Goldstein Marine Mammal Research and Rehabilitation Center with a hospital for dolphins, whales, and sea turtles, and the Keating Marine Education Center. The Goldstein Marine Mammal Center also houses exhibits for turtles, manatees, and dolphins, and a newly constructed marine turtle exhibit sponsored by Penelope Kingman.

The gradual metamorphosis of Mote from a small field station to a major research center is more than I ever dreamed—thanks to our great team of scientists, administrators, volunteers, and supporters who share the joys, excitement, and challenges of working at a place seeking knowledge of the sea.

Forty years ago I wrote, in the then last chapter of this book, about two very important people who took Mote giant strides forward. Perry Gilbert, the brilliant zoologist, who I knew could handle the growth of the Lab from a field station to a major research center, and William R. Mote, the successful business man who loved fishing and wanted to further the study of the sea. Through research and understanding, Bill Mote wished to return to the sea the bountiful gifts it has given us. These two were the most important of our many great supporters. I proudly salute them for their amazing accomplishments. For the last 25 years, we have been very fortunate that Dr. Kumar Mahadevan, from India, has taken the reins as President and CEO. The many members of our extended "Mote family" include a large friendly group of staff, visiting scientists, students, interns and volunteers.

My personal family—my four children, Hera, Aya, Tak and Niki, now in their fifties, and my grandson, 20 year old Eli Weiss, are all scuba divers and frequently accompany me on our research expeditions.

Hera (Norma Sophia Konstantinou) and her husband, Coz, have been happily married 25 years and have a lovely home in Florida. Both studied marine science in graduate school at Texas A&M but, for the last 12 years, have been involved in dog agility for their six retrievers. They are especially proud of their two top-ranked agility dogs that they rescued from shelters.

"Aya" (Ayame Iris Yumico Maria Konstantinou) spends more time in the air than underwater, flying everything from Boeing 727s to 767s as a Captain for Continental Airlines. She compares flying

Hera and Coz of OceanWolf Agility with their agility champion, Kai.

around clouds to floating suspended over coral reefs. Her son, Eli, now 20, attends University in Sarasota. He is a champion at BMX racing, and designs and builds his own BMX tracks that were featured in *Popular Mechanics*. He started scuba diving at age five. Two of his photographs of whale sharks, taken when he was five, were featured in *National Geographic* in 1997. He is the youngest photographer ever published by *National Geographic*.

Both my sons are single at present. "Tak" (Themistokles Alexander Konstantinou) works as a real estate agent for Michael Saunders in Sarasota. In elementary school and high school, he was a champion on the swim teams; and he continues to swim competitively. Tak loves photography, and has won prestigious awards in Sarasota for his artistic photographs. His precision photographs of the fishes I am studying, both topside and underwater, have been valuable in my studies and our scientific publications.

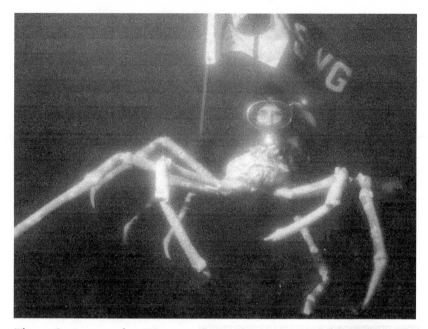

The Society of Women Geographers had just awarded Genie their prestigious gold medal, previously awarded to few others including Amelia Earhart and Margaret Meade. The SWG asked Genie for her picture with a marine animal. The giant crab at a depth of 140' tried to carry Genie away by wrapping his hind legs around her thighs.

photo credit David Doubilet

My grandson, Eli Weiss, at age 5, was the youngest photographer featured in National Geographic in June 1997, with his photos of a whale shark he took while snorkeling.

Investigating the sleeping sharks of Mexico. David Doubilet took remarkable photographs of my graduate student, Anita George, daughter Aya, and me.

After spending 20 years in Kauai as a dive shop operator and underwater photographer, Niki, (Nikolas Masatomo Konstantinou) my youngest, has been working for ten years for Cirque du Soleil's "O" as a safety diver below its water stage in Las Vegas. He has achieved the Shodan level (black belt) in Iaido, and he comes to Sarasota every December to give a classical piano recital for our annual get together. In October 2011, we went to Japan, to see Niki's class demonstrate Iaido in Himeji, to view the magnificent Okinawa Churaumi Aquarium where they have kept as many a five whale sharks at the same time, and I was able to introduce my daughter, Aya, to the Emperor where they talked more about tennis than we usually talk about fish.

Over the years, members of my family, Mote volunteers, and volunteer divers from all over the world have accompanied me on 117 field studies sponsored by the University of Maryland Foundation and more than 20 other expeditions supported by the National Geographic Foundation, Office of Naval Research, National Science Foundation, etc. Expeditions to study sand-dwelling fishes, coral reef fishes, and sharks have been made to Indonesia, Papua New Guinea, Malaysia, Thailand, Solomon Islands, Mexico, the Caribbean, and the Red Sea. My first expedition, supported by a Fulbright Scholarship, was to Egypt and the Red Sea. This expedition helped to launch much of my later research and expeditions.

Helen Vanderbilt, one of our earliest supporters, was in her 60s when she learned to scuba dive because she wanted to join our expeditions. Like me, she was amazed at the beauty and tremendous abundance of marine life at Ras Mohammed in the Red Sea at the tip of the Sinai Peninsula. She backed me to the hilt and lobbied with me to protect this area and try to make it into a national park. President

Stan Waterman a dear friend and diving buddy of my family.

The Mote Campus, City Island, Sarasota. Left front the Keating Education Building, right front Aquarium and Laboratory buildings, rear left the Goldstein Marine Mammal Hospital.

Sadat's son, Gamal, introduced me to his father. President Sadat was surprised to learn that Egypt had such a beautiful underwater area and agreed to make it a national park by presidential decree. He had one condition, that when Gamal dived with me at Ras Mohammed, I protect him from the "maneating" sharks. I was preparing to attend the ceremony when President Sadat was to declare Ras Mohammed

285

Thousands of baby convict fish swarm over me as I carry the Wings World Quest flag.

a national park when we learned he had been assassinated. His presidential decrees were discarded and I thought all was lost in our endeavor. I later found out that the only way to make Ras Mohammed a permanent national park was to have it declared so by the Egyptian parliament. Helen Vanderbilt and I worked hard to do this. A year later our lobbying efforts paid off and Egypt had its first national park, Ras Mohammed—the first underwater park in the Red Sea. Even the governor of South Sinai was talked into being a supporter. He asked if he could have a statue of himself placed at the entrance of the park. He later asked me, in private, "But why do we need such a large area to park cars?"

The success of our many diving expeditions is due to the talents and expertise of many remarkable dive teams and trained observers. Some have participated in more than 20 expeditions. Diane Nelson, a professor at East Tennessee State University, leaves her studies of *tardigrades* (microscopic "water bears") to dive with us and coauthor

On my 87ᵗʰ birthday, I was offered a test drive in the new Aviator,
a two person diving submarine, by Captain Alfred McLaren.

fish manuscripts. I sometimes escape to the beautiful Blue Ridge Mountain home of Drs. Diane and Jack where we focus on writing up our data. John Pohle, a retired military colonel, applies his meteorological, mapping, and organizational skills to our expeditions. His late wife Mary, our videographer and coauthor of one of our scientific papers on sandfishes of New Guinea, was an important part of our team. She was a beautiful person who gave up a noted career in figure skating and joined our underwater research on fishes. Dr. Ben Kendall, a medical doctor, and his artist wife, Ginny, are another important couple in our diving team. We are always entertained by their closing night hilarious songs and cartoons summarizing each of our expeditions. We also get free medical advice. Tom Alburn and his wife, Patrice Boeke, have created some of our most innovative equipment for counting and studying small fish. Patrice timed the hatching of *Trichonotus* eggs in glass goblets in our ship's dining room that were collected in the sands after the dawn matings of *Trichonotus*.

Ruth Petzold, one of our prize underwater photographers, switches her artificial land leg to her dive leg, and gets up at 5 a.m. to get her

Best friend for 82 years, Norma Denman Woodburn and Genie holding book showing them as teenagers.

extraordinary photos of sandfishes mating at dawn. A countess from Spain, Maya Moltzer, is an extraordinary videographer who looks like a mermaid underwater documenting our studies. Mary Jane Stoll, another prized videographer, explores the Indo-Pacific and has led us to many of our best dive sites. Don Blair and Patty Gergen have also documented our underwater studies with valuable photographs. David Shen has recorded and photographed the rare behaviors of new species of fishes, and documented so much of our work. He has had two new species of fish named after him by diving scientists and noted icthyologists, Dr. John Randall and Dr. Wolfgang Klausewitz, and David's photos have appeared on the cover of scientific publications. Dr. Steve Kogge, both a photographer and a computer whiz, designs some of the specialized equipment we use in our research dives. Like a coral reef proctologist, he made a remarkable video inside the convict fish tunnels. His video documents the behavior of the adult convict

Coz (son-in-law), Hera (daughter), Niki (son),
Genie, Tak (son),Aya (daughter)

fish, who hide all day in their winding branching tunnels under the coral reef and are fed by the swarms of babies who return to the tunnels each night after feeding on plankton. The thousands of young convict fish form a giant swarm resembling a marine mammal turning into

*Our second dive trip to Indonesia on the **Seven Seas**, September - October 2009. Crew and research divers studying Trichonotus. Back row- Steve Kogge, Madi Verbeek, Greg Hoffmann, John Pohle, 4 crew, Jann Rosen-Queralt, Captain Wahyudin Ismail, Aya Konstantinou, crew; 3rd row - Tak Konstantinou, Maya Moltzer, Pat Shaw, Martha Kiser, Mary Jane Stoll, Jack Nelson, Karl Klingeler; 2nd row - Ruth Petzold, Alice McNulty, Cathy Marine, Diane Nelson, Eugenie Clark; 1st row - six crew members of the Seven Seas.*

➤
Trichonotus elegans *males defending the territorial boundary between their harems.*

290

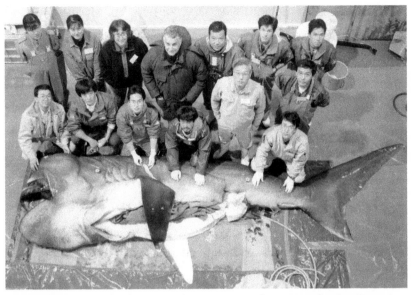

José Castro, center back row, and I were invited to Japan for the dissection of the first female megamouth shark ever found. Team leader Kazuhiro Nakaya, second from right front row, and Kazunari Yano to his right.

photo credit Senzo Uchida

a monstrous sea snake before they disappear into a small hole for the nightly feeding of their cryptic parents.

In the fall of 2009, my children Aya and Tak, plus twenty-two other research divers, including Mote divers Cathy Marine, Amy Fleischer, Brad Tanner, and Greg Hoffmann and a high school scholar, Madi Verbeek, accompanied me to Indonesia for a month of diving to study the behavior of large colonies of trichonotids (a type of sandfish). We have studied these sandfishes in the Red Sea off and on since 1960. The sandfishes, protogynous hermaphrodites, start as smaller drab females and change to colorful flamboyant males. They live in harems on the sand when not hiding during the day from possible enemies and "sleeping" at night. The family Trichonotidae has only about six species, one of which is *Trichonotus nikii*, named after my son, Niki. *T. nikii* lives mainly in the white sand of the Red Sea, where we have studied rare large populations concentrated in four areas of the Gulf of Aqaba. Each of these colonies is fed by "nutrient chutes," water or wind-driven enriched sand that constantly feed the colony. *Trichonotus elegans* is a more widely distributed species that we have studied in a few other places in the world. Sometimes swarms of hundreds or thousands of *T. elegans* feed together, very similar to *T. nikii.*

The other four or five species of *Trichonotus* (aside from the Red Sea *T. nikii* and the Indo-Pacific *T. elegans*) I call "perchers." Instead of swarming above their well-defined territories on the sand, they feed mainly on benthic organisms. The individuals are larger than the swarming species. We named a new species of perchers, *Trichonotus halstead* from Observation Point in Papua New Guinea, after Bob and Dinah Halstead, the greatest diving guides and photographers in that part of the world.

In Indonesia, we worked primarily around the island of Sangeang, northwest of Komodo National Park. Sangeang has an active volcano that periodically sends down lava creating "nutrient chutes" to an area on the western side of the island where we found swarms of *T. elegans* thriving over and on and in the relatively black sand. A "temporary" village, Bontoh, has a small population and about 300 of their water buffalo roam near the shore, defecating and adding additional nutrients to the area, which supports huge colonies of garden eels mixed with *T. elegans*, similar to those seen at Ras Mohammed in the Red Sea.

I consider myself to be extraordinarily lucky to have had so many wonderful friends, most of whom are somehow associated with my fun and fascinating work with fishes. My closest friend since fourth

Eugenie and Tak in Mexico when they worked on sleeping sharks

grade, Norma Woodburn, comes to visit me several times a year from her home in Plymouth, Massachusetts. She and her daughter, Lee, have participated in our expeditions to Papua New Guinea, Belize, Borneo and Mexico. During her Florida "vacations," Norma works full time with me at Mote.

In "Shark Alley" at Mote, I am surrounded by good friends in my field who share my love for fishes. Across the hall from my office is Dr. José Castro, a worldwide authority on sharks. He has just completed a monumental new book, published mid-2011, *Sharks of North America*. Together, we sometimes dissect very large sharks at Mote's Marine

Mammal Necropsy Room (see photo of big-eye thresher dissection on p.276). In the office next to mine is my colleague, Dr. Carl Luer, who went to school with my children, and is now an authority on the physiology and biochemistry of sharks, skates, and rays. He and Dr. Cathy Walsh, another "Shark Alley" neighbor, study the immune system of these fishes, which have a remarkable resistance to cancer. Out most recent addition to "Shark Alley" is Dr. Nick Whitney, who is studying the hydrodynamics of swimming movements. Heading our shark group is Dr. Bob Hueter, who is currently studying in the Caribbean the largest known feeding aggregation of whale sharks in the world—well over 1,000 individuals.

Other ichthyologist friends are as far away as the other side of the world. Recently, on the way back from a diving expedition to Papua New Guinea, I revisited the Japanese Emperor and Empress during a plane stop in Tokyo. His Majesty always sends his chauffeured limousine to pick me up at my hotel and bring me through the heavily guarded, beautiful palace grounds. We enjoy sharing the latest in our fish research. I showed him our remarkable video of the swarming young convict fish returning to their parents, and the Emperor told me about his most recent studies of the DNA and behavior of hover gobies, such as *Ptereleotris*. The hover gobies are also plankton feeders, and hide like *Bathygobius* in sand tunnels on the sea floor. His descriptions brought back memories of our first search for gobies at a Miami beach, where we found them hiding in empty beer cans (chapter 14)! The courtesy of the Emperor and Empress escorting me out to the limousine for the return to my hotel is a special memory.

Other distant fish enthusiasts I consider very special friends include the late Dr. Che Tsung Chen, who was the leading shark expert in Taiwan and president of National Taiwan Ocean University.

He introduced me to the whale shark fishermen in Taiwan, which led to the discovery of a whale shark carrying 300 embryos, the largest number of embryos found in any shark. Senzo Uchida was the first to keep whale sharks alive for many years in his gigantic glass windowed aquariums on Okinawa, the latest being the Churaumi Aquarium. He invited me to watch the sharks' behavior and special feeding method for these plankton eaters. The sharks come to the surface by the feeding platform to slurp the mass of plankton ladled into the water. They move their heads from side to side to get the expanding concentrate of plankton. Through the kind hospitality of Drs. Yasuo Suyehiro and Hajime Masuda, I was able to meet other Japanese biologists and divers who have inspired and helped me in my shark studies: Dr. Tokiharu Abe, the late Dr. Kazunari Yano (one of my former students), Dr. Kazuhiro Nakaya of Hokkaido University, as well as the great divers/photographers Koji Nakamura and Yasumasa Kobayashi.

Dr. Hamed Gohar, director of the marine biological station at Ghardaqa, was my co-worker and the co-author of my first large paper on Red Sea fishes. Ayman Taher, an outstanding Egyptian diver and photographer as well as Nick Caloyianis (my former student and teaching assistant) and his partner Clarita Berger in Caloyianis Productions worked with the Egyptian Telmassani brothers to produce the movie "Treasures of Ras Mohammed" helping promote Ras Mohammed as an area to be preserved. David Fridman was the Curator of Fishes of the Red Sea in two aquariums including Coral World. He spent more time underwater and was the greatest naturalist and expert of Red Sea fishes, and knew how to exhibit them in extraordinary ways to show their beauty and fantasy-like qualities. Until his death a few years ago, he was the most sought after consultant for Red Sea marine life. Howard Rosenstein was another great Red Sea diver and leader of expeditions. I learned so much from the great Israeli marine scientists Dr. Lev Fishelson, Dr. Adam Ben-Tuvia, and Dr. Heinz Steinitz who enabled me to take part in their Red Sea expeditions.

Rodney Fox developed the cages for viewing sharks underwater and is the white shark expert of Australia. Ramon Bravo, who is Mexico's "Jacques Cousteau" and is a diver and cinematographer, discovered the Cave of Sleeping Sharks and invited me to dive there. Jacques Cousteau invited me to be guest scientist on the first shark episode in the series "The Underwater World of Jacques Cousteau." We were honored to have him serve on the Cape Haze Marine Laboratory International Advisory Committee along with Dr. Yasuo Suyehiro of Japan, who visited our lab in its early stages and arranged my scientific tour of Japan and my first meeting with the Crown Prince Akihito, before he became Emperor.

Other special friends include my severest and best critic, Dr. Vic Springer of the Smithsonian in Washington, DC and Dr. Jack Randall of the Bishop Museum in Honolulu, who holds the world record in number of new fish species (680) described. I cannot come even close to completing this list of special friends without mentioning Dr. Jack Garrick, who was a Professor of Zoology at Victoria University at Wellington and the leading shark expert of New Zealand. Nearly 30 years ago, he arranged for me to get extraordinary deep-sea specimens of sharks and a chimera with seven claspers for teaching labs in my graduate classes in marine vertebrates at the University of Maryland. Shark specimens for these labs were also provided by local Maryland fishermen. Without the kindness and expertise of these scientists and divers and fishermen, I could not have enjoyed studying so many unusual fishes during my lifetime.

In 2009 I celebrated my 87th birthday by learning to pilot a deep-diving submersible called 'Super Aviator' and then 'fly' it deep under the surface of Nevada's Lake Tahoe. This sub ride was followed by a surprise party organized by Captain Alfred McLaren with many of my sub-diving friends.

In 2010 I turned 88, an important year and number in several cultures, including Japanese. Following my 88th birthday, I had the thrill of diving for the first time with large seven-gill sharks in the kelp beds off Cape Town, South Africa, courtesy of the Save Our Seas Foundation, where I now serve as a scientific advisor. In April 2012, on a research expedition to study sand fishes in Indonesia, I celebrated my 90th birthday aboard the MV *Aurora* with long-time diving friends.

Cover photo for the October 2011 issue of the scientific journal, aqua. Our study of the poisonous catfish, Plotosus, was enhanced by the marvelous photographs of my diving friend, Yasumasa Kobayashi. He captured, for the first time, their remarkable cleaning behavior.

Our current research on ocean triggerfish, tricky sand fishes, and garden eels still takes us to warm tropical seas. After the work and fun of expeditions, we publish our observations to share with the scientific community. In January 2011 Jack Randall and I described a new species of swellshark, *Cephaloscyllium stevensi*, found in deepwater nautilus traps off Papua New Guinea. The manuscript on observations in the Indo-Pacific of the venomous catfish, *Plotosus*, was the cover article for the October 2011 issue of *aqua: International Journal of Ichthyology*. I am thankful for the help of my numerous coauthors, especially Diane Nelson and John Pohle, and the editorial help of many, including Bev Rodgerson, Lance Ong, and Rachel Dreyer.

I hope to continue learning about fish for many more years to come. Then maybe I'll meet up with Dr. Breder and we could resume our long talks about the wonderful ways of fishes.

Bibliography

CHAPTER 1: DISCOVERING A WATERY EDEN

Bigelow, HB, Schroeder WC. 1948. *Fishes of the Western North Atlantic, vol. 1: Sharks.* New Haven: Sears Foundation for Marine Research. 576 p.

Heller, JH. 1960. *Of Mice, Men and Molecules.* New York: Scribner. 176 p. (Contains a chapter on his shark work at Cape Haze Marine Lab.)

Heller, JH, Heller, MS, Springer, S., Clark, E. 1957. Squalene content of various shark livers. *Nature* 179: 919-20.

CHAPTER 3: SHARKS AND ABDOMINAL PORES

Carrier, JC, Musick, JA, Heithaus, MR, editors. 2004. *Biology of sharks and their relatives.* Boca Raton: CRC Press. 596 p.

Daniel, JF. 1922. *The Elasmobranch Fishes.* Berkeley: University of California Press. 332 p. (on shark anatomy)

Gilbert, PW, Wood, FG. 1957. Method for anesthetizing large sharks and rays safely and rapidly. *Science* 126: 212-3.

Hyman, LH. 1992. *Hyman's comparative vertebrate anatomy.* 3rd ed (originally published 1979). Chicago: University of Chicago Press. 788 p.

CHAPTER 4: THE MYSTERY OF SERRANUS

Atz, JW. 1964. Intersexuality in fishes. In: Armstrong, CN, Marshall, AJ, editors. *Intersexuality in Vertebrates.* New York: Academic Press. p 145-232.

Breder, CM, Jr., Rosen, DE. 1966. *Modes of Reproduction in Fishes.* Garden City, N.Y.: Natural History Press. 941 p.

Clark, E. 1959. Functional hermaphroditism and self-fertilization in a serranid fish. *Science* 130: 215-6.

Clark, E. 1965. Mating of groupers. *Natural History* 74: 22-5.

Harrington, RW Jr. 1961. Oviparous hermaphroditic fish with internal fertilization. *Science* 134: 1749-50.

Mead, GW. 1960. Hermaphroditism in archbenthic and pelagic fishes of the order Iniomi. *Deep Sea Research* 6: 234-5.

CHAPTER 5: BLENNIES AND MAZES

Fowler, HW. 1954. Description of a new blennioid fish from southwest Florida. *Notulae Naturae* No. 265: 1-3.

Tavolga, WN. 1954. A new species of fish of the genus *Blennius* from Florida. *Copeia* No.2: 135-9.

CHAPTER 6: SHARKS THAT RING BELLS

Clark, E. 1959. Instrumental conditioning of lemon sharks. *Science* 130: 217-8.

Clark, E. 1962. The maintenance of sharks in captivity. Part 1, General. *Bulletin Institute de l'Oceanography, Monaco.* Special No. 1A: 7-13.

Springer, S. 1948. Oviphagous embryos of the sand shark, *Carcharias taurus. Copeia* No. 3: 153-7.

Springer, S. 1950. Natural history notes on the lemon shark, *Negaprion brevirostris. Texas Journal of Science* No. 3: 349-359.

Springer, S. 1960. Natural history of the sandbar shark, *Eulamia milberti. U.S. Fish and Wildlife Service Fishery Bulletin* 61(178): 1-38.

CHAPTER 7: THE SHARK HAZARD

Clark, E. 1960. Four cases of shark attacks on the west coast of Florida, Summer 1958. *Copeia* No. 1: 63-7.

Coppelsdon, VM. 1958. *Shark Attack.* Sydney: Angus and Robertson Ltd. 219 p.

Davies, DL. 1964. *About Sharks and Shark Attacks.* Pietermaritzburg, S. Africa: Shuter and Shooter. 237 p.

Gilbert, PW, editor. 1963. *Sharks and Survival.* Boston: D. C. Heath & Co. 578 p.

Gilbert, PW. 1962. The behavior of sharks. *Scientific American* 207 (July): 60-8.

Gilbert, PW, Mathewson, RF, D. P. Rall, DP, editors. 1967. *Sharks, Skates, and Rays.* Baltimore: Johns Hopkins Press. 623 p.McCormick, HW, Allen, TB, Young, WF. 1963. *Shadows in the Sea: The Sharks, Skates and Rays.* New York: Chilton Books. 415 p.

Young, WF. 1934. *Shark, Shark: The 30-year odyssey of a pioneer shark hunter.* New York: Gotham House. 287 p.

CHAPTER 8: SCIENTISTS AND STUDENTS

Sudak, FN. 1966. Immobilization of large sharks by means of ethanol. *Copeia* No. 3: 611-2.

Wright, T, Jackson, R. 1964. Instrumental conditioning of young sharks. *Copeia* No. 2: 409-12.

CHAPTER 9: MESOPLODON

Moore, JC. 1960. New Records of the Gulf Stream beaked whale, *Mesoplodon gervaisi,* and some taxonomic considerations. *American Museum Novitates* No. 1993. 1-35.

CHAPTER 10: SINKHOLES AND "RAPTURE OF THE DEEP"

Burgess, RF. 1976. *The cave divers.* New York: Dodd Mead. 239 p.

Clausen, CJ, Brooks, HK, Wesolowsky, AB. 1975. Florida spring confirmed as 10,000 year old early man site. *Florida Anthropologist* 28(3): Part 2. 38 p.

Oakley, KP. 1960. Ancient preserved brains. *Man* (London) 60: 90-1.

Purdy, BA. 1991. *Art and archaeology of Florida's wetlands.* Boca Raton: CRC Press. 317 p.

Royal, W, Clark, E. 1960. Natural preservation of human brain, Warm Mineral Springs, Florida. *American Antiquity* 26: 285-7.

CHAPTER 11: MORE EDUCATED SHARKS

Clark, E. 1962. Visual discrimination in the lemon shark. *Tenth Pacific Science Congress, Honolulu, Hawaii, 1961. Symposium Abstracts.* 175-6.

Clark, E, von Schmidt, K. 1965. Sharks of the central west coast of Florida. *Bulletin of Marine Science.* 15: 13-83.

Gruber, SH, Hamasaki, DH, Bridges, CDB. 1963. Cones in the retina of the lemon shark (*Negaprion brevirostris*). *Vision Research*

3:797-8.

Hodgson, ES, Mathewson, RF. 1978. *Sensory biology of sharks, skates, and rays.* Arlington, VA: Office of Naval Research, Department of the Navy. 666 p.

Hueter, RE, Cohen, JL, editors. 1990. Vision in elasmobranchs: A comparative and ecological perspective. *The Journal of Experimental Zoology* (Supplement 5) 5: 1-182.

Hueter, RE, Mann, DA, Maruska, KP, Sisneros, JA, Demski, LS. 2004. Sensory biology of elasmobranchs. In: Carrier, JC, Musick, JA, Heithaus, MR, editors. *Biology of sharks and their relatives.* Boca Raton: CRC Press. p 325-368.

Sperry, RW, Clark, E. 1949. Interocular transfer of visual discrimination habits in a teleost fish. *Physiological Zoology* 22: 372-378.

CHAPTER 13: CHILDREN AND TRAVELS

Clark, E, von Schmidt, K. 1966. A new species of *Trichonotur* (Pisces, Trichodontidae) from the Red Sea. *Sea Fisheries Research Station, Haifa.* 42:29-36.

CHAPTER 14: AN IMPERIAL ICHTHYOLOGIST

Aronson, LR, Aronson, FR, Clark, E. 1967. Instrumental conditioning and light-dark discrimination in young nurse sharks. *Bulletin of Marine Science* 17: 249-256.

CHAPTER 15: MANTAS AND OTHER RAYS

Clark, E. 1963. Massive aggregations of large rays and sharks in and near Sarasota, Florida. *Zoologica* 48: 61-64.

Coles, RJ. 1916. Natural history notes on the devilfish, *Manta birostris* (Walbaum) and *Mobula elfersi* (Muller). *Bulletin of the American Museum of Natural History* 35: 649-57.

Roosevelt. T. 1917. Harpooning devilfish. *Scribner's Magazine* 62: 293-305.

CHAPTER 17: FISHY ADVENTURES CONTINUE

Balon, EK. 1994. The life and work of Eugenie Clark: devoted to diving and science. *Environmental Biology of Fishes* 41: 89-114.

Castro, JI. 2010. *The Sharks of North America*. New York: Oxford University Press.

Clark, E. 1975. Into the lairs of "sleeping" sharks. *National Geographic* 147: 570-584.

Clark, E. 1982. Secrets of the Red Sea. *Science Digest* 90(4): 46-53.

Clark, E. 1983. Hidden life of an undersea desert. *National Geographic* 164(1): 128-144.

Clark, E. 1992. Gifted guidance to Egypt's wondrous reefs. *Sea Frontiers* 38(5) 20-27.

Clark, E, Kogge, SN, Nelson, DR, Alburn, TK, Pohle, JF. 2006. Burrow distribution and diel behavior of the coral reef fish *Pholidichthys leucotaenia* (Pholidichthyidae). *aqua, International Journal of Ichthyology* 12: 45-82.

Doubilet, AL. 2008. Quest for the poisonous catfish. *The Explorer's Journal* 86: 46-50.

Doubilet, AL. 2008. Shark Lady: In conversation with Eugenie Clark. *The Explorer's Journal* 86: 52-53.

Hueter, R, Gonzalez Cano, J, De la Parra, R, Tyminski, J, Perez-Ramirez, J, Remolina-Suarez, F. 2007. Biological studies of large feeding aggregations of whale sharks (*Rhincodon typus*) in the southern Gulf of Mexico. In: Irvine, TR, Keesing, JK. editors. The First International Whale Shark Conference: Promoting International Collaboration in Whale Shark Conservation, Science and Management. Conference Overview, Abstracts and Supplementary Proceedings. CSIRO Marine and Atmospheric Research, Australia. p 76.

Hueter, RE, Tyminski, JP, De la Parra, R. 2008. The geographical movements of whale sharks tagged with pop-up archival satellite tags of Quintana Roo, Mexico. Second International Whale Shark Conference, Holbox, Quintana Roo, Mexico. http://www.domino.conanp.gob.mx/doc_conf/Bob.pdf.

Joung, S-J, Chen, C-T, Clark, E, Uchida, S, Huang, WYP. 1996. The whale shark, *Rhincodon typus*, is a livebearer: 300 embryos found in one 'megamamma' supreme. *Environmental Biology of Fishes* 46: 219-223.

Luer, CA, Walsh, CJ, Bodine, AB. 2004. The immune system of sharks, skates, and rays. In: Carrier, JC, Musick, JA, Heithaus,

MR, editors. *Biology of sharks and their relatives.* Boca Raton: CRC Press. p 369-395.

Motta, PJ, Maslanka, M, Hueter, RE, Davis, RL, De la Parra, R, Mulvaney, SL, Habegger, ML, Strother, JA, Mara, KR, Gardiner, JM, Tyminski, JP, Zeigler, LD. 2010. Feeding anatomy, filter-feeding rate, and diet of whale sharks *Rhincodon typus* during surface ram-filter feeding off the Yucatan Peninsula, Mexico. *Zoology* (in press).

Vanderbilt, HC, Clark, E. 1983. A scientific and conceptual endeavor for the development of coastal marine parks and mariculture. *Bulletin of the Institute of Oceanography and Fisheries (Cairo)* 9: 477-478.

about EUGENIE CLARK

Mote Marine Laboratory's founding director, Dr. Eugenie Clark, received the Explorers Club Medal, March 15, 2008 at age 85. The award is the highest honor bestowed by the club, recognizing extraordinary lifetime contributions in exploration and scientific research. Genie, as she is known, is an ichthyologist who began her studies of fish as a child visiting the Battery Park Aquarium in New York City. She wanted to grow up and study fish, a rare goal for a girl. She began her research on the behavior and reproductive isolating mechanisms of fresh-water aquarium fishes but later combined her love for diving with the study of marine fishes – first by hard-hat diving and snorkeling, and then using scuba and submersibles. She has searched for sharks in submersibles diving to depths of 12,000 feet. She has headed diving expeditions to the Caribbean, Bermuda, the Pacific and Indian Oceans, Indonesia, Papua New Guinea, the Red Sea and Japan to study the behavior of sharks and fishes of the sand/coral reef and continues to do so.

Dr. Clark has a B.A. from Hunter College, New York; M.A. and Ph.D. degrees from New York University; and three Honorary D.Sc. degrees from the University of Massachusetts, the University of Guelph and Long Island University. Her career began as a research assistant at Scripps Institution of Oceanography, at the New York Zoological Society, and at the American Museum of Natural History in New York. She was the founding director (1955 to 1967) of the Cape Haze Marine Laboratory, now the Mote Marine Laboratory in Sarasota, Florida, a leading center for shark research, red tide, aquaculture, etc. In 1968, she became a professor at the University of Maryland, College Park, in the Department of Biology, taught there for 32 years and is now a Professor Emerita. Currently she is back at Mote Marine Laboratory, as Founding Director and Senior Research Scientist.

Dr. Clark is the recipient of seven fellowships and scholarships, seven medals (including the gold medal of the Society of Woman Geographers), and 36 other awards for work in marine biology, conservation, exploration, teaching and writing. She is one of the world's authorities on sharks, and author of more than 170 scientific articles and popular books on sharks and other fishes. She has carried the flag of the Society of Woman Geographers to Ethiopia and underwater off Japan and Egypt; in 1980 she carried the flag of the National Geographic Society to Egypt, Israel, Australia, Japan, and Mexico; in 1986 it was the flag of the Explorers Club on the "First American Diving Expedition to South China" and in 1988 she carried the flags of all three to the bottom of Monterey Canyon (12,000 ft).

She received a Fulbright Scholarship in 1951 to spend a year studying fishes of the Red Sea at Egypt's marine biological laboratory at Ghadaqua. This led to 44 more trips to both the Egyptian and Israeli coasts of the north Red Sea and Dahlak in the south Red Sea. She was the first foreigner to become an honorary member of the Zoological Society of Israel. Dr. Clark's efforts led to the creation of

*Genie at her 88th birthday party with her children Niki, Aya and Tak,
and her shark cake and convict fish cupcakes*

the 1st National Park (underwater) of Egypt, Ras Mohammad, and
the story of her Red Sea exploration, told in *Lady with a Spear,* led
to her opportunity to establish the Cape Haze Marine Laboratory.

Among her many other honors, she received the "Cousteau
Award" for underwater science from the Boston Sea Rovers; the
Key to the City of Sarasota with her birthday, May 4, established
as Eugenie Clark Day, and the Henry Luce Award of $10,000 from
WINGS World Quest. In 2010 she was inducted into the International
Scuba Diving Hall of Fame in the Cayman Islands, received the
National Council of Jewish Women "women in power" award, was

inducted into the Florida Governors Hall of Fame, and received the Lifetime Achievement Award from the government of Bonaire. Fellow scientists have named four fish after her: *Callogobius clarki* (a goby named by Menachen Goren); *Sticharium clarkae* (a blenny by Anita George & Victor G. Springer); *Enneapterygius clarkae* (a barred triplefin by Wouter Holleman); and *Atrobucca geniae* (a Red Sea snapper by Adam Ben-Tuvia & Ethelwynn Trewavas).

She continues to explore the unknown in the sea and write scientific papers on fishes, and has managed a successful and amazing career while raising 4 children, a role model for both men and women who are inspired to become marine biologists and scientists. As she said at the Explorers Club: "I can't believe that I've been given this medal for having such a lifetime of fun diving and studying fish."

INDEX

314

CPSIA information can be obtained
at www.ICGtesting.com
Printed in the USA
BVHW081001301218
536679BV00001B/69/P

9 781936 051526